Portraits of the Earth

A Mathematician
Looks at Maps

Mathematical World
Volume 18

Portraits of the Earth

A Mathematician
Looks at Maps

Timothy G. Feeman

AMERICAN MATHEMATICAL SOCIETY

2000 *Mathematics Subject Classification*. Primary 00–01, 26A06, 51M09, 86A30; Secondary 00A69, 51–01, 51M25, 86–04.

Library of Congress Cataloging-in-Publication Data

Feeman, Timothy G., 1956–
   Portraits of the earth : a mathematician looks at maps / Timothy G. Feeman.
      p. cm. — (Mathematical world, ISSN 1055-9426 ; v. 18)
   Includes bibliographical references and index.
   ISBN 0-8218-3255-7 (softcover : alk. paper)
   1. Mathematical geography.   2. Cartography.   I. Title.   II. Series.

GA13 .F44   2002
526—dc21   2002027950

Dedicated to the memory of Elaine Bosowski.
The joy, inspiration, and love she brought to everyone she knew
and everything she did remain vibrant and alive.

Contents

Preface

Migratory animals such as whales, caribou, sea turtles, butterflies, and birds seem to have built-in maps that enable them to travel between precise locations without any apparent external assistance. Humans have also migrated considerably, from our origins in Africa to Asia, Oceania, Europe, and the Americas, and have otherwise travelled extensively for thousands of years. Evidence of maps used to describe our surroundings, to interpret the natural environment, to orient ourselves, and to aid in navigation dates back as far as our knowledge of human civilization. Although we can't say for certain when the first maps were made or for what specific purpose, it seems likely that some of the earliest maps were used to tell where a given something or somewhere was located – a productive fishing spot, fresh water, or a dangerous place to be avoided.

Today we encounter maps on a daily basis. Maps from the simple and sublime to the complex and confusing. From the symbols on restroom doors to street maps of the towns and cities where we live, from shopping mall directories to detailed geologic maps depicting the underground structure of the earth, from traffic signs to world maps accompanying newspaper articles about global events – all are maps. They guide and direct our daily routine activity; they help us to make difficult decisions. They can be used to present information, to formulate policies, or to explore. Whatever its purpose, every map is a tool, a product of human effort and creativity, that represents some aspects of our world or universe. Every map is a representation, an approximation, a reduction of something that is larger or more complex.

In order to make its point, every map actually distorts the truth, distorts reality, by employing some type of code to formulate and express its message. For many maps, one ingredient of this code is a mathematical transformation of the map's subject. This is most readily apparent, perhaps, in the sorts of maps shown in a typical atlas, but it is true of many other maps as well. In this book, we will focus on world maps, maps that portray relatively large portions of the earth's surface. We will study the different attributes that maps can have and determine mathematically how to construct maps that have the features we want. We will see how maps are used to communicate and how the distortions that arise in making maps of the world affect our ability to communicate. We will use maps to explore and to navigate across the seas and through the air. Along the way, we will learn of the longstanding collaboration and interrelationship between geography and mathematics. We will travel to every continent, visit civilizations both ancient and contemporary, and meet a few of the many outstanding individuals who have influenced the way we look at and map our world.

This book grew out of an interdisciplinary undergraduate course in mathematics and cartography designed by the author and the late Dr. Elaine F. Bosowski at

Villanova University. The course was powered by our belief that, by exploring the mathematical ideas involved in creating and analyzing maps, students would see how mathematics could help them to understand and explain their world. At the same time, by actually creating maps of their own, students could develop the skills needed to accurately evaluate the quality of graphic images they encounter on a routine basis. In other words, the interface between mathematics and cartography seemed to us to be an exciting and accessible area of study for college students – well, we were excited by it anyway! We called our course *Cartographiometry*, developed it with support from the National Science Foundation's Mathematics Across The Curriculum program (grant # NSF-DUE-95-52464), and offered it to students at Villanova University during the 1996-97 academic year. In the class, the students explored, as we will do in this book, the shape of the earth, the determination of latitude and longitude, elementary spherical geometry, the uses and computation of scale factors, the design of optimal routes for air or sea navigation, Gaussian curvature, the reasons why we can't make a perfect flat map of the earth, how to evaluate from a critical and analytical viewpoint maps that we encounter every day, and how to design atlas maps using both hand-drawn techniques and computer graphics software. Since Dr. Bosowski's untimely death in 1998, variants of the *Cartographiometry* course have been taught in a distance-learning format and as a senior mathematics seminar.

In this book, the intent is not to give a historical overview of humankind's attempt to map our world, but to investigate in some detail the mathematics involved in doing so. Necessarily, the discussion will become somewhat technical at times as we work through the analysis and equations needed for one map or another. This may seem a bit tedious in places, but there is no escaping it other than to accept someone else's word on how to design a map. If we want to understand this process and be able to construct and analyze maps for ourselves, we must pursue the mathematical details.

Despite its more or less obvious presence, the mathematics behind various maps is rarely discussed in any undergraduate mathematics or geography course, except, possibly, as an interesting side topic in a first course in differential geometry taken by advanced mathematics and physics students (cf. [McCleary, 1994]). Some books on map projections written by cartographers (e.g., [Pearson, 1990], [Bugayevskiy and Snyder, 1995]) take this approach as well, though few undergraduate students in geography have the mathematical prerequisites to use them. Other books, such as the classic text of Deetz and Adams, present tables of calculations indicating how to construct various maps while essentially suppressing the actual mathematics. Most general cartography textbooks (see [Dent, 1996] and [Robinson, 1995]) include a brief overview of map projections and their properties with no attempt at mathematical analysis. Of course, as both the earth and the flat piece of paper onto which it is to be mapped are two-dimensional surfaces, a differential geometry approach, together with the background in multivariable calculus it presupposes, might seem to be necessary for understanding maps. In fact, this is not the case. As we shall see here, an important key to understanding certain properties of projections, such as the distortion of areas or angles, is an analysis of a map's scale factors. It turns out that calculating scale factors involves only the comparison of linear distances, so the basic elements of one-variable calculus usually will be adequate to our task.

The principal mathematical tools we will use are basic calculus – derivatives and integrals – and trigonometry. Many of the maps we will study were originally developed centuries or even millenia ago, and there is a significant gap in the levels of familiarity with trigonometry between mathematicians of the nineteenth century and earlier and we mathematical thinkers of today with our handy computational machines. So we should not be surprised if we must dust off some half-forgotten trigonometric relationships in the course of our investigations. A few sections of the book draw on the concept of vectors in two- or three-dimensional space, and one chapter (on obliquely centered maps) employs three-by-three matrices and a bit of linear algebra as central tools. You will miss some interesting ideas, but still get the main gist, if you skip those sections.

Many of the world maps that you see throughout the text were generated using the *Maptools package* created during the spring of 2001 by Vincent Costanzo, who was a student at Villanova University at the time. The package, which uses a database of coastal points developed by the United States Geological Survey, works within the computer algebra system Maple and is available at the Maple applications center website.

I owe an immeasurable debt to Elaine Bosowski, late of the Villanova University Geography Department. This book would not have been written without the collaboration we formed in the years before her death. I would also like to thank the students who took the *Cartographiometry* course in any of its various forms; the National Science Foundation, and especially Lee Zia, for supporting the *Cartographiometry* project; all those colleagues who responded with ideas and encouragement to various talks I have given on this subject; Gary Michalek for reading the whole manuscript carefully; Bob Jantzen for generously sharing his knowledge of LATEX and Maple; and Ed Dunne, at the American Mathematical Society, for his invaluable support and advice. I take full responsibility for any shortcomings and mistakes that remain and welcome comments from readers.

Villanova University, Villanova, Pennsylvania, USA
75° 20′ 31.0″ W; 40° 2′ 24.2″ N.
timothy.feeman@villanova.edu

CHAPTER 1

Geodesy – measuring the earth

[In world cartography] the first thing that one has to investigate is the earth's shape, size, and position with respect to its surroundings, so that it will be possible to speak of its known part, how large it is and what it is like ... .

These things belong to the loftiest and loveliest of intellectual pursuits, namely to exhibit to human understanding through mathematics ... [the nature of] the earth through a portrait ... . (Claudius Ptolemy, [Berggren and Jones, 2000])

Before we can make maps of the world, we are called upon to determine the size and shape of the earth, establish the location of points on it, and measure the shapes and areas of various parts of its surface. These studies comprise the subject of **geodesy**, a close relative of the field of geometry. Much of the mathematics we will discuss in this book will involve applying our knowledge of geometry and geodesy to cartography.

1. The shape of the earth

As Ptolemy suggested, the fundamental question of geodesy is "What are the shape and size of the earth?" The notion that the earth is essentially spherical dates back at least to the sixth century B.C., when Anaximander and Thales of Miletus, two of the earliest classical Greek geometers, recorded their beliefs that the earth was a sphere positioned at the center of a huge celestial sphere to which the other visible planets and stars were fixed. Pythagoras, in the fifth century B.C., further propounded this idea, and Aristotle and Plato philosophized that the earth, home to humanity, must have been perfectly created and, hence, must have the perfect geometric shape – a sphere. (See [Cotter, 1966].) The image of the earth as a sphere has persisted, though not without some significant lapses. In recent times, photographs of our planet taken from space have reinforced this image.

In the late seventeenth century, the great physicist and mathematician, Isaac Newton, suggested, on the basis of his work on gravitation and planetary motion, that the shape of the earth could be more appropriately described as an *oblate spheroid* or *ellipsoid* — not a perfect sphere, but a sphere-like shape that is oblate, or flattened, at the poles and bulges somewhat south of the equator. A survey conducted in France by the Cassinis, father and son, between 1684 and 1710, gave further evidence that the earth was not a true sphere, though they apparently believed that the earth might be elongated rather than flattened at the poles. Subsequent expeditions to Peru and Lapland during the 1730's, which measured arcs of meridians at high and low latitudes, confirmed Newton's belief that the earth has the shape of an oblate spheroid. In subsequent years, meridian measurements were made at numerous locations in order to determine the earth's true shape. Since the introduction of satellite technology in the latter part of the twentieth century, new measurements of the earth have resulted in the development of several "reference ellipsoids". These include the World Geodetic Systems ellipsoid of 1984 (WGS 84, for short), developed by the US Defense Mapping Agency, and the Geodetic Reference System ellipsoid of 1980 (GRS 80), adopted by the International Union of Geodesy and Geophysics in 1979.

Of course, if one thinks about it, the earth can't *really* be either a sphere or an oblate spheroid. These are smooth, mathematically defined surfaces, without mountains, valleys, canyons, oceans, ice caps, or shifting continents. The term *geoid*, which simply means *earthlike*, is used to define the shape that most closely approximates the shape of the earth with all its bumps, bulges, depressions and divisions. The exact shape of the geoid is designed to match the earth's gravitational field and average sea level at each location. There are both physical and computer models of the geoid that have been carefully formed to represent the earth's shape, but they are of limited use in mapmaking.

For mapping on a global scale, we usually adopt a regular shape to approximate the earth's surface, because of the scale at which we are working. When the geoid is generalized, smoothed out, or averaged to form a more regular shape, of a size we can easily handle, the result turns out to be only negligibly different from a sphere. So the sphere is, for many purposes, a satisfactory approximation of the earth's shape. In fact, the ridges found on most globes at seams in the globe's construction are far more exaggerated than the tallest of mountains and deepest of ocean ridges would appear on a globe of such small dimension. When the highest level of precision is needed, an oblate spheroid provides the best model for the earth's shape. However, using this model adds technical difficulties to the mapping process, as we shall see in Chapter 12. In practice, a flat map of the ellipsoid can be made by composing one of the many map projections available for the sphere with a map of the spheroid onto the sphere.

In this book, we assume a spherical earth. All of the measurements and calculations we make will be done assuming the earth to be a sphere.

2. The size of the earth

Assuming the earth to be spherical, the single most important measurement to be made is to determine its circumference, and, consequently, its radius. This was precisely the task undertaken over two thousand years ago, *circa* 230 B.C., by Eratosthenes, who lived, worked, and studied in Alexandria and was the first person, as far as we know, to scientifically calculate and record the size of a planet, the earth.

Eratosthenes realized that at noon, when the sun is highest in the sky, on any given day of the year two sticks, called *gnomons*, of the same length would cast shadows of different lengths when stuck in the ground at two different places located north and south of one another. Specifically, he knew that at noon on the summer solstice, usually June 21, the sun shone directly down a well in the town of Syene, now the site of Aswan, in Egypt. A gnomon placed in the ground at Syene would cast *no shadow* at noon on that day, but a gnomon placed in the ground at Alexandria, located north of Syene at the mouth of the Nile River, *would* cast a

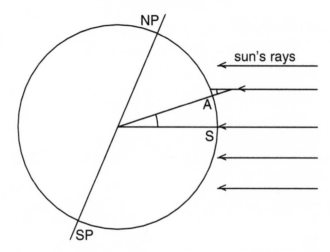

FIGURE 1. The sun, overhead at Syene (S), makes an angle with a gnomon at Alexandria (A).

shadow. Assuming the earth to be a sphere and all of the sun's rays to be parallel (a reasonable assumption given how far the sun is from the earth), Eratosthenes called upon a basic geometric fact — that the interior angles on the same side of a line transversal to two parallel lines must add up to 180°. Figure 1 illustrates this geometry. Taking the line from the center of the earth through Syene and out to the sun as one of the parallel lines, the line of the sun's rays in Alexandria as the other parallel line, and the line from the center of the earth out to the tip of the gnomon in Alexandria as the transversal, Eratosthenes was able to conclude that the angle between Syene and Alexandria, measured at the center of the earth, would be the same as the angle between the gnomon in Alexandria and a string connecting the tip of the gnomon to the tip of its shadow. On the appointed day at the appointed time, Eratosthenes determined this angle to be one-fiftieth part of a full circle. It followed that the distance between Syene and Alexandria must be one-fiftieth part of a full circumference of the earth! It remained only to measure the distance between the two towns. This was probably achieved using camels and general knowledge at the time about the pace of these animals. In any case, the distance was determined to be about 5000 stadia (a common unit of the time and roughly equivalent to one-tenth of a mile). Multiplying by 50 yields Eratosthenes' estimate of 250000 stadia (roughly 25000 miles) for the circumference of the earth. This is remarkably accurate given the methods used to measure angles and distances as well as the problem that Syene and Alexandria are not exactly due north and south of each other. The basic method employed by Eratosthenes is completely valid and is still used today.

Table 1 gives measurements for the equatorial, polar, and mean (average) diameter and circumference of the earth. These are different due to the oblate spheroidal shape of the earth. Assuming the earth to be a sphere, use whichever of these figures seems most appropriate for the context. For instance, to measure the distance from Beijing to the north pole, use the polar circumference. For two places near the equator, use the equatorial distance. In general, use the mean measurement.

	size of the earth	
	diameter	circumference
equatorial	7926.38 miles	24901.46 miles
	12761.47 km	40071.02 km
polar	7899.80 miles	24855.34 miles
	12718.68 km	39936.65 km
mean	7917.52 miles	24861.01 miles
	12747.21 km	40026.24 km

TABLE 1. Approximate equatorial, polar, and mean dimensions for the earth

3. Latitude

The earth rotates in space about an axis through its center. The ends of the axis are called the north and south poles. The imaginary circle on the earth's surface that lies halfway between the two poles is called the *equator*. The equator divides the earth into two equal pieces called the *northern* and *southern hemispheres*. The circumference of the equator is the same as that of the sphere.

As Eratosthenes observed, the farther north one goes from Syene on the summer solstice, the lower the sun will be in the sky at noon. So a gnomon will cast a longer shadow the farther north it is placed. It follows that we can determine how far north and south two places are from each other by comparing the angles of the sun at noon on any given day. This is the concept known as *latitude*. Officially, the latitude of any point on earth is equal to the difference between the angle made by the sun at noon at the point in question and the angle made by the sun at noon of the same day at the equator. This is the same as the angle of inclination of the point away from the plane of the equator, as illustrated in Figure 2. Latitude is designated as north (N) or south (S) according to which hemisphere the point lies in. The latitude at San Francisco, California, is 38° N. This means that San Francisco is north of the equator and that at noon on any given day the sun is 38 degrees lower in the sky at San Francisco than at the equator. The latitude of the north pole is 90° N, while the south pole is located at latitude 90° S. The equator itself has latitude 0.

The poles are the only points at their respective latitudes, but in general each latitude is shared by many points all at the same angle north or south relative to the equator. These points form an imaginary circle on the earth's surface called a *parallel* because it lies in a plane parallel to the plane of the equator. Thus, San Francisco lies on the parallel at 38° N.

Four parallels, in addition to the equator, have special importance for the earth as a planet. We know that the earth revolves around the sun and that the earth's axis is tilted in relation to the plane of the earth's orbit. The path of the earth's orbit is called the *ecliptic* and the plane that it defines is referred to as *the plane of the ecliptic*. The earth's axis makes an angle of about 23° 30′ (23 degrees, 30 minutes, or twenty-three and one-half degrees) with the plane of the ecliptic. As the earth goes through its orbit, there are two days each year when an observer standing on the sun would see the earth's axis tilting completely to the right or to the left and not toward the sun at all. On these two days at noon the sun will shine

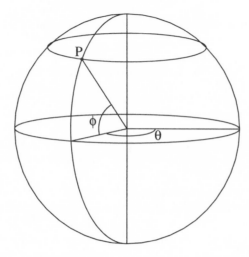

FIGURE 2. The point P has longitude θ and latitude ϕ.

directly overhead at any point on the equator. Day and night will be exactly equal in length at every point on earth. These two days are the *vernal* and *autumnal equinoxes* and occur approximately on March 21 and September 20. On two other days each year, an observer at the sun would see the earth's axis tilting entirely toward or away from the sun, and not at all to the right or left. On one of these days at noon the sun will shine directly overhead at any point on the parallel at latitude $23° 30'$ north, and on the other the sun will shine directly overhead at noon along the parallel at latitude $23° 30'$ south. These days are known as the *summer* and *winter solstices*, respectively, and occur approximately on June 21 and December 20. The parallel at latitude $23° 30'$ N is known as the *Tropic of Cancer* while the parallel at latitude $23° 30'$ S is called the *Tropic of Capricorn*. To remember these, you might think about which constellations of stars are visible in the night sky on these dates to an observer in the northern hemisphere.

On the summer solstice, when the midday sun shines directly overhead at the Tropic of Cancer, places whose latitude is more than 90 degrees away from the Tropic can't see the sun because they are tilted too far away from it. Since $90 - 23.5 = 66.5$, we see that places at latitudes between $66.5°$ and $90°$ south will have twenty-four hours of darkness on the summer solstice. By the same token, between $66.5°$ and $90°$ north there will be twenty-four hours of daylight on the summer solstice. These situations are reversed on the winter solstice. The parallels at $66° 30'$ N and $66° 30'$ S are called the *Arctic Circle* and the *Antarctic Circle*, respectively.

The four special days we have just discussed divide the year into seasons. Winter solstice, when the north pole is tilted farthest from the sun, marks the start of winter in the northern hemisphere. The onset of spring is marked by the vernal equinox which is followed by the summer solstice bringing summer to the northern hemisphere. Finally the autumnal equinox marks the beginning of the autumn or fall season, and then the cycle repeats itself. The seasons are opposite in the southern hemisphere where summer starts in December and autumn in March. The names of the solstices and equinoxes are a sign of northern hemisphere, European bias. Remember that north and south are names created by humans. There is no "up" in space!

4. Longitude and time

As the earth travels along its orbit around the sun, it also rotates about its axis. As the earth rotates, the position of the sun in the sky changes. We see the sun appear on the eastern horizon, then rise in the sky until it is at its zenith, or highest position, then descend until it sets in the west and disappears from our view. Thus, the position of the sun in the sky tells us how far around we are in the earth's rotation.

The moment at which the sun reaches its zenith is called *local solar noon*. This event occurs simultaneously at all points along a semi-circular arc, called a *meridian*, that extends from the north pole to the south pole. Half-way around the world from this meridian is another meridian experiencing local midnight, exactly half a day away from local noon. Where two meridians come together at the poles, they form an angle that is the basis for determining *longitude*. **Latitude and longitude together give a complete system for locating points on the earth.** But how do we determine our longitude?

5. Determining longitude – two problems

The equator, lying half-way between the poles, forms a natural reference point for measuring latitude. There is no such convenient reference for measuring longitude. The way the day progresses is pretty much the same from one meridian to the next. How do we choose one meridian, with its local noon, to be special? Just as importantly, once we choose one meridian as a reference, how do we figure out how far around our meridian is from that one? The answer to the first question is political, to the second celestial and mechanical. Let's tackle the second question first.

5.1. Your noon, my noon. Suppose you and I are on different meridians and that we wish to determine the angle between our two meridians where they come together at the poles. We know that it takes 24 hours for the earth to rotate about its axis once. So if my local noon was to occur 4 hours after yours, say, then I must be one-sixth of the way around the world from you. There are 360 degrees in a full circle, so one-sixth of that is 60 degrees. Our meridians would form an angle of 60 degrees. If we were to choose your meridian as the special reference meridian, so your longitude would be automatically 0, then my longitude would be 60 degrees west. Why west and not east? Because my local noon occurred four hours *after* yours, not four hours before.

This is all well and good, until we realize that in order for me to know that my noon came four hours after yours, I would have had to know when your noon occurred. There are two basic ways to solve this problem. One is celestial. For instance, suppose that a solar eclipse took place one day exactly at local noon in your location. I would see the solar eclipse take place at exactly the same moment and note that this event took place four hours before my local noon. If we were to share our notes about the solar eclipse, we would see that my local noon was four hours after yours. In fact, Christopher Columbus was able to use an occurrence of a lunar eclipse to aid him in determining his longitude. The relative positions in the sky of various celestial bodies, such as certain stars, the planets, or the moons of Jupiter, have also been widely used to determine longitude. Tools such as the astrolabe and the sextant were developed to measure these positions. The Global Positioning System (GPS) devices of today use signals from satellites, in place of the celestial bodies, to compute longitude. To measure longitude using celestial events or the positions in the sky of celestial bodies generally requires communication with an observer located at a reference meridian, elaborate tables of known positions of celestial objects, or sophisticated computer technology. For instance, to compute longitude while at sea during the mid-eighteenth century, a ship's navigator would take measurements of Jupiter's moons using difficult-to-manage instruments and then spend several hours making computations.

A mechanical approach to determining longitude is based directly on measuring time for, if we think about it a bit more, we might realize that what we really need is a dependable clock that tells the local time at the reference meridian. Sea voyagers, for instance, would periodically reckon their own local time during their journey. The difference between the time on the reference clock and the voyagers' local time would show how far east or west they had travelled. Of course, the reference clock has to be accurate and able to stand up to the rigors of a sea voyage.

Though sufficiently accurate clocks are readily available today, their introduction just a few centuries ago marked a giant leap in the technology of navigation. Without the ability to determine longitude at sea, ships would often sail along a parallel, making them easy prey for pirates. Many ships simply got lost or ran aground due to miscalculation of their position. The first sea-worthy chronometer was developed in England by John Harrison and presented to the English Longitude Board in 1735. It took some years and further refinements, especially in size, weight, and ease of reproduction, for the new clocks to catch on. The book *Longitude*, by Dava Sobel, presents a fascinating account of Harrison's travails and the quest to determine longitude.

5.2. Ready for prime time? Knowing how to determine relative longitudes of different locations still leaves the problem of choosing one meridian as a special reference, the so-called *Prime Meridian*. Ptolemy placed his Prime Meridian through the Canary Islands off the west coast of Africa. Others have used the meridians through Mecca, Jerusalem, Paris, Rome, Copenhagen, the Cape Verde Islands, St. Petersburg (Russia), Philadelphia (U.S.A.), and more. When the Royal Observatory in Greenwich, England, published, in 1767, the most comprehensive tables of lunar positions available, sailors the world over found themselves calculating their longitude and setting their chronometers from Greenwich. This common practice became official in 1884 when the International Meridian Conference established the Prime Meridian at Greenwich, where, incidentally, John Harrison's original clock, weighing seventy-five pounds and measuring nearly four feet across, is still running smoothly.

5.3. The International Date Line. Longitude is measured as an angle east (E) or west (W) from the Prime Meridian. Exactly half-way around, where 180° west seems indistinguishable from 180° east, a sort of "discontinuity" in time occurs. The clocks at 180° E run 12 hours *ahead* of the clocks in Greenwich while those at 180° W run 12 hours *behind*. Neighbors have clocks that are off by nearly 24

hours! It is here that the International Date Line has been established. It is the
line where today meets yesterday and tomorrow meets today.

In fact, the International Date Line is not exactly the meridian 180 degrees
away from Greenwich as that really would result in neighbors, though not too
many of them, living in different days. Instead, the Date Line makes a few zigs and
zags while generally staying near the 180° east/west meridian.

5.4. What time is it where you live? Each location has its own local solar
noon that is used to compute its exact longitude. However, it is inconvenient for
the scheduling of events, such as airplane departures and starting times of movies,
to have people use clocks set to their own local time. After all, local time changes
as soon as one moves to the east or west even the slightest amount! Time zones
were created to address this problem. In the United States this occurred, largely
on the initiative of the railway companies, on November 18, 1883, which became
known as "the day of two noons". As accustomed as we are to time zones now,
it is important to remember the distinction between the official time of our time
zone and the local time determined by the position of the sun. It is the latter that
determines our longitude.

6. A word on coordinates

The method of locating points on a sphere by determining the appropriate angles of longitude and latitude constitutes a basic coordinate system on the sphere.
Mathematicians often use a similar framework, called *spherical coordinates*, for locating points in three-dimensional space. In mathematical spherical coordinates,
one first determines the distance from a given point of interest to some predesig-
nated origin. In practice, we don't need this number to help us locate points on
the earth's surface because the earth has a fixed radius. Also, in mathematical
spherical coordinates, latitude, usually denoted by ϕ, is measured as an angle away
from the north pole. We will stick with the geographers' convention of measuring
latitude from the equator. The angle denoted by θ in spherical coordinates measures the amount of rotation about the vertical axis. This is the same as longitude.
Where cartographers usually measure angles in degrees, mathematicians use radians in order to simplify trigonometric computations. In this book, **for computations involving longitude and latitude, we will measure angles in radians.
Also, we will use positive and negative angle measurements instead of
the directional designations north/south or east/west, with south and
west getting assigned negative values. We will use the symbol θ to denote longitude and ϕ to denote latitude.** Thus, the point with longitude 75°
W and latitude 40° N will be assigned coordinates ($\theta = -75\pi/180$, $\phi = 40\pi/180$)
while the point with longitude 75° E and latitude 40° S will be assigned coordinates
($\theta = 75\pi/180$, $\phi = -40\pi/180$). The north pole has latitude $\pi/2$, the south pole
has latitude $-\pi/2$, and all points on the equator have latitude 0. The International
Date Line has longitude $\pm\pi$.

Occasionally, it will be more convenient to use the *Cartesian coordinate system*
in three-dimensional space. In this case, we will take as the z-axis the line through
the north and south poles, with the north pole on the positive z-axis. The plane of
the equator corresponds to the xy-plane with the positive x-axis meeting the equator at the prime meridian and the positive y-axis meeting the equator at the point

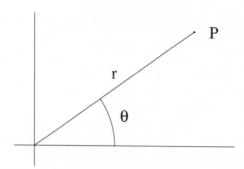

FIGURE 3. The point P has polar coordinates (r, θ). Its Cartesian coordinates are $x = r\cos(\theta)$ and $y = r\sin(\theta)$.

with longitude $\pi/2$ (or $90°$ E). On a sphere of radius R, then, the point with longitude θ and latitude ϕ (in radians with the $\pm$ conventions just discussed) will have Cartesian coordinates $(x, y, z) = (R\cos(\theta)\cos(\phi), R\sin(\theta)\cos(\phi), R\sin(\phi))$. Conversely, one can recover the latitude and longitude from the Cartesian coordinates by observing first that $\phi = \arcsin(z/R)$. This determines ϕ uniquely as an angle between $-\pi/2$ and $\pi/2$. Also, observe that θ satisfies the equations $\tan(\theta) = y/x$, $\cos(\theta) = x/\sqrt{x^2 + y^2}$, and $\sin(\theta) = y/\sqrt{x^2 + y^2}$. Any two of these together will determine a unique angle θ between $-\pi$ and π.

For flat maps of the earth, we will use two different coordinate systems in the plane – Cartesian and polar.

The two-dimensional Cartesian coordinate system uses the familiar perpendicular x and y axes with the origin at the point of intersection. The axes are oriented so that the positive y-axis is at an angle of $\pi/2$ radians measured counter-clockwise from the positive x-axis.

The *polar coordinate system* in the plane also utilizes two components to designate location – a distance, usually denoted by r, and an angle, generally denoted by θ. Begin by choosing an origin, which can be any convenient point in the plane. Next, draw a half-line that emanates from the origin and extends infinitely. Any other half-line that emanates from the origin then defines an angle between $-\pi$ and π radians measured counterclockwise (or, for negative angles, clockwise, really) from the reference half-line. Each point in the plane is located by a pair (r, θ), such as $(3, \pi/4)$, in which the first number, r, is the distance from the point in question to the origin and the second, θ, is the angle between the reference half-line and the half-line that emanates from the origin and passes through the point. See Figure 3. To facilitate the conversion of polar coordinates to Cartesian coordinates, and vice versa, it is customary to align the positive x-axis so that it coincides with the half-line corresponding to $\theta = 0$. Then the point with polar coordinates (r, θ) has Cartesian coordinates $x = r\cos(\theta)$ and $y = r\sin(\theta)$. Moreover, $x^2 + y^2 = r^2$ and, when $x \neq 0$, $\theta = \arctan(y/x)$.

7. Exercises

(1) Write down the locations, in terms of latitude and longitude, of five different places on earth, no two of which are on the same continent, and describe fully what is located at each place. Determine the latitude and

longitude of the antipode of your own current location and describe what is located there. Write down the name and location of a place which will not see the sun on December 21st. Write down the name and location of a place at which the sun passes directly overhead on December 21st.

(2) The circumference of the parallel at latitude ϕ on the earth is $2\pi R \cos(\phi)$, where R is the radius of the earth.

   (a) Use the *mean diameter* of the earth from Table 1 to determine the circumferences, in miles and in kilometers, of the parallels at latitudes 15, 30, 45, 60, and 75 degrees (north or south).

   (b) At latitude 40° North, how far west of Philadelphia would you have to travel to reach a place whose local time is one hour behind Philadelphia's local time?

   (c) Use the *equatorial diameter* of the earth from Table 1 to determine the length, in feet and in meters, of a *nautical mile*, defined to be 1/60 of a degree of longitude measured at the equator.

(3) At latitude 20° S, a train leaves station A at *local time* 10:15 in the morning. The train travels due east at a constant speed of 100 kilometers per hour until it reaches station B, 25 kilometers away. What is the *local time* at station B when the train arrives there?

CHAPTER 2

Map projections

The design of a world map, or even a map of a small part of the world, necessarily involves transferring certain measurements from the earth to the map, such as the land area of Australia or the distance from Tokyo to the north pole. We cannot merely copy the true measurements (unless we intend to make a map the size of the earth itself!) but rather must formulate rules for scaling them down. As we shall see, the rules we select have a lot to do with the final appearance and useful properties of the maps we make.

Cartographers refer to any systematic representation of the earth's surface onto another surface as a **projection**. Thus, when we make a map of the world on a surface other than the globe, we are constructing a map projection.

1. Three classes of projections

Taking the term "projection" literally, we might construct a map by laying a spherical globe on a flat sheet of paper, with the south pole at the bottom say, and placing a light source at the center of the sphere. Suppose the globe is transparent, with a grid of selected meridians and parallels (called the **graticule** by cartographers) drawn on its surface. As the light shines through the globe, the grid's shadow will be projected onto the paper. The shadows of the meridians and parallels become the basis for a map. Figure 1 illustrates this basic construction.

In this case, the south pole will be its own shadow in the center of the paper and the meridians will project as straight lines radiating out from there. As we move from the south pole up to the equator, the parallels will project onto larger and

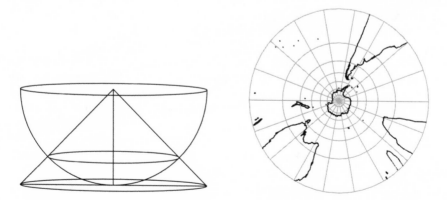

FIGURE 1. The gnomonic projection. The basic construction is depicted on the left, and the resulting map of most of the southern hemisphere on the right.

larger circles increasingly far apart. The equator and everything above the equator
will cast no shadow at all on the paper. Thus, we will get a map of only half of
the globe and with considerable distortion away from the south pole, as shown in
Figure 1. This particular map projection, one of the oldest known to humans, is
called a *gnomonic projection*. Points other than the south pole can be used as the
center and we shall discuss later the purpose in that.

The gnomonic projection is an example of a *planar* or *azimuthal* projection, so
named because the paper onto which the projection is made lies flat in the form of
a plane. Other azimuthal projections can be created by placing the light source at a
different point along the axis of the globe. For instance, the *stereographic projection*
uses a light that is placed at the point on the globe opposite to where the paper
touches the globe, and the *orthographic projection* is formed by placing the light
at 'infinity' so that the light's rays are all perpendicular to the sheet of paper.
Azimuthal projections can also be defined purely mathematically rather than by
any projection in the literal sense. The general format is of a central point (one of
the poles) with the meridians radiating out and the parallels shown as concentric
circles about the central point. Of course, it is not possible to draw *all* of the
meridians and parallels, so, in practice, we select them at regular increments of
10, 15, or perhaps 20 degrees. For instance, the expression "a 15 degree graticule"
means that the meridians and parallels are depicted every 15 degrees.

Another approach to creating a map projection is to begin by forming a sheet
of paper into a cylinder that fits around the globe, touching it at the equator say.
As before, the grid of meridians and parallels on the globe can be projected out
onto the cylinder which in turn can be slit open and laid out flat to produce a map.
A map formed in this way is called a *cylindrical projection*.

For instance, we might again imagine that there is a light source located at
the center of a transparent globe and look at the shadow cast by the graticule on
the surrounding cylinder. The sphere and cylinder touch along the equator so the
equator is its own shadow. The meridians cast long vertical shadows and, hence,
are mapped as evenly spaced vertical lines on the paper. All parallels have circular
shadows that wrap around the cylinder. The parallels get smaller toward the north
and south poles, but their shadows always have the same circumference as the
cylinder. This has the effect of greatly elongating the east-to-west dimension of
regions near the poles. Also, in this scenario, the projections of the parallels are
spaced farther and farther apart as they get nearer to the poles. Therefore, polar
regions appear enlarged north-to-south as well.

A different cylindrical projection, shown in Figure 2, results if we dispense with
the light source and map points on the sphere horizontally out to the nearest point
on the cylinder. Again, the equator maps onto itself and, again, the meridians are
mapped as evenly spaced vertical lines. Also, parallels are projected onto circles of
the same circumference as the cylinder so that non-equatorial regions appear elongated east-to-west on the map. Unlike with the previous projection, the images
of the parallels now get closer together on the map as we move toward the poles.
This is because the images of successive parallels would be separated only by the
vertical distance between the parallels on the globe. As we move toward the poles,
this vertical separation shrinks so that the north-to-south dimensions become compressed on the map. Thus, in this purely mathematical scenario, polar regions will
appear shorter and wider than they actually are. This particular projection was

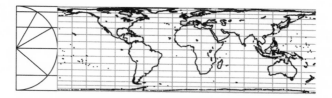

FIGURE 2. Lambert's equal-area cylindrical projection

introduced by Johann Heinrich Lambert in 1772 and, as we shall discuss later, has the property that the areas of all regions of the globe are shown in their correct proportions.

In general, a cylindrical projection has the form of a rectangular grid with the meridians, selected at regular intervals, depicted as evenly spaced vertical lines or line segments and the parallels, also selected at regular increments, as horizontal segments all having the same breadth as the equator.

A third class of projections are the *conic projections* in which we place a cone on the globe and somehow project the points on the globe onto the cone. The cone is then slit open and laid out to form a sector of a circle. Conic projections were used by Ptolemy *circa* 150 A.D. and are especially useful for mapping portions of the globe that are wide east-to-west but short north-to-south (such as the United States). The United States Geological Survey uses conic projections extensively in creating its official topographical maps. Both cylindrical and conic projections can be modified to have two lines where the sphere and the cylinder or cone meet.

Some of the maps we will discuss, such as those of Mercator and Mollweide, are not projections at all in the literal sense, though cartographers nonetheless refer to them as such, but are purely mathematical constructions designed to have certain attributes. Also, a map need not be conveniently centered on one of the poles or on the equator, as in the illustrations above. For instance, you might prefer a map centered on your own city or your own meridian. A map that is centered on a pole is said to have a *polar perspective* while a map centered on the equator has an *equatorial perspective*. The perspective of a map centered at any other location is referred to as *oblique*. As we shall see, map analysis is generally easier if one adopts a polar or equatorial perspective.

2. Map attributes

We shall see later on that there can be no ideal flat maps of even a portion of a sphere. That is, no flat map can show all distances in all directions in their correct proportions. We shall also see that no flat map of a sphere can both show all shortest routes between arbitrary points on the sphere as straight line segments *and*, at the same time, preserve all angles between routes. Therefore, at least one of these features of the globe must become distorted on the map. If angles are distorted, then shapes and directions will be altered. If shapes or angles are preserved, then distances must be distorted, so areas will be altered. In other words, no matter how hard we try, something has got to give and any map we make will incorporate significant distortions of the original spherical world.

In analyzing the properties of any given map projection or in trying to design a map having certain prescribed properties, we will focus on trying to measure or control the distortions in the following features.

- *Distance/Shortest routes.* To what extent does the map represent true distances between places on the earth? Does the shortest path between two points on the earth show up as a straight line segment on the flat map?
- *Area/Size.* Are regions of the earth shown on the map with the correct proportional areas?
- *Angle/Shape.* Are angles on the sphere portrayed accurately on the map? Does the map show the true shapes of regions on the earth's surface?
- *Direction.* Are paths of constant compass bearing shown as straight lines on the map?
- *Visual appearance.* Does the map reasonably convey the appearance of the earth?

It is possible to make a map that exhibits any one of the first three properties individually, but always at the expense of distortions in the other two features.

A map that portrays *all* distances accurately is called an *isometry*. As we shall discuss later, Euler proved, in 1775, that it is not possible to construct an isometry from a sphere onto a flat surface. It *is* possible to design a flat map of the earth that preserves all distances from a single point to any other point on the globe. Such maps are called *equidistant* projections. The gnomonic projection, introduced above and discussed in more detail later, always maps the shortest routes between points on the sphere onto straight line segments.

A map that depicts the areas of all regions of the globe in their correct proportions is called an *equal-area* projection. However, to show areas in their correct proportions in all cases forces regions to be compressed in some directions and simultaneously stretched in others, thus distorting both shapes and distances.

A map that preserves angles and shapes at every location is called a *conformal* map. A conformal map that shows the equator and the meridians as straight lines therefore requires an expansion of the graticule at the poles because the meridian lines must be shown as perpendicular to the line of the equator and hence parallel to each other. This expansion of the polar regions automatically creates distortions in distance and area. There are conformal maps, such as the stereographic projection, that show either the equator or the meridians or both as curved lines but these nonetheless distort distances, directions, and areas.

Finally, maps that preserve direction have been developed as tools to aid in navigation. The most famous of these types of maps, developed during the mid-sixteenth century by Gerhard Mercator, shows all lines of constant compass bearing as straight lines. This map is also conformal, so, as just mentioned, it distorts areas and distances especially in the polar regions.

As we examine map projections in greater detail, we will study how each of these attributes can be achieved in a suitably constructed map.

3. Attributes and map use

No single map projection of the entire world is perfect for all purposes. A map that accurately shows distances to a central point can help us to judge the time or amount of fuel needed to travel to or from that place. An equal area map would be appropriate if we wished to produce a world map that displayed area-based data. Equal area projections have also been at the heart of recent debates over what some have seen as the eurocentric influence of certain maps that originated during the periods of European dominance and colonialism. A conformal map most accurately depicts how things look on the globe, but only locally because of distortions in area over large regions. Important landmarks and their relative dimensions and orientations are more easily recognized on conformal maps making them valuable for navigation. A map that preserves direction helps us to get where we want to go, to navigate effectively. Finally, the choice of a central point or perspective for a map determines which parts of the world will undergo the least (and the most) distortion.

By learning how to construct various map projections, by learning to identify a map's properties based on mathematical analysis, we also learn about the limitations and appropriate uses of maps. We learn to appreciate the value and power of maps of all types.

4. Map projections *vs.* mappaemundi

The earliest uses of projections to depict the earth go back to classical Greece and Egypt. The gnomonic projection was apparently known to Thales of Miletus in the sixth century B.C. while Ptolemy of Alexandria, in the second century A.D., constructed several influential map projections in detail. Between Ptolemy's time and the onset of the renaissance in Europe, however, there were few developments in the area of map projections. Some notable exceptions include the work of Islamic scholars such as al-Biruni around 1000 A.D.

Many of the maps that were developed in Europe during the Middle Ages were based not on any projection but on a philosophical view of the world. Among these were world maps, or *mappaemundi*, that portrayed the earth's surface symbolically, often based on a literal interpretation of the Bible. One particular type of *mappamundi* that emerged in Europe during the sixth and seventh centuries, when the works of Eratosthenes and Ptolemy were either unknown, lost, unaccepted, or even viewed as heretical, was the **T-in-O** map that schematically portrayed the biblical division of the world's lands among the three sons of Noah.

A T-in-O map is so named because the general form of the map, depicted in Figure 3, can be represented by a letter "T", representing the Mediterranean Sea, inscribed in a letter "O", representing the great world ocean. Early explorers assumed that it was possible to travel between the world's lands on this one world ocean, a belief that, while faulty, nevertheless led to grand voyages of discovery. The world's three known major land masses, Africa, Europe, and Asia, are separated by the "T" and the "O". (Of course, the Americas were as yet unknown to Europeans.) Beyond this basic format, many T-in-O maps showed detailed coastlines, navigable rivers, and ports. Some were even used for navigation purposes in local areas. On most, the center of the map is focused on Jerusalem as the "cradle of civilization", the center of world faiths and a pivot point of global understanding at the time. Also, since the known world (at least to Europeans) was centered around the Mediterranean Sea, and Asia was the least known (again, to Europeans) among the world's land masses, it was thought that "Paradise" must lie in the far reaches of unknown Asia. This notion was furthered by the discovery of the Spice Islands in Indonesia and the idea that these islands might actually be Paradise. Thus,

FIGURE 3. A rudimentary T-in-O map showing the Mediterranean
Sea as a letter "T" inscribed in a letter "O", representing the great
world ocean.

"Paradise", when drawn, was depicted far off, beyond Asia, at the very top of the
T-in-O map.

   Although the T-in-O map form was both commonly accepted and widely used
in its time, the desire for knowledge and the demands of exploration and navigation
eventually led to the resurgence of a scientific approach to mapmaking in Europe.

CHAPTER 3

Scale factors

We humans are quite small compared to the size of the planet we inhabit. Even a fairly small part of the earth's surface is too large for our vision to encompass. Thus, the maps we make are not only representations of the earth, with certain features exaggerated and others ignored, but representations *on a different size scale* than the planet itself. Maps are like scale models of the actual object they portray. The number called the *scale factor* measures how the sizes of things depicted on the map, specifically distances between points, compare to the sizes of the real objects on the earth.

1. Indicating map scale factors

For a scale model of an object, we think of the scale factor as a ratio – the size of the model divided by the actual size of the object it represents. More accurately, the scale factor measures the degree to which *each dimension* of the object has been reduced in size. For instance, for a $\frac{1}{100}$ scale model automobile, the real car would be 100 times as long, 100 times as wide, and 100 times as high as the model. In cartography, the scale of a map is called the *RF* or *representative fraction* and is indicated on the map either as a fraction, such as $\frac{1}{1\,000\,000}$, or as a comparison, such as 1:1 000 000, with the colon representing the word "to." Both of these *RF*'s would be read as "one to one million", a phrase that itself constitutes an example of a *verbal scale*, a scale given in words. The representative fraction for a map does not depend on the unit of measurement one uses, but, rather, should be interpreted as meaning that one unit of any length on the map represents one million (or whatever the appropriate value is) of the *same units* measured on the surface of the earth. Typical *RF* values are 1:40 000 000 for an atlas map of South America, 1:12 000 000 for a map of the United States, and 1:100 000 for a map of a large city.

The scale of a map can also be indicated with a *graphic* or *bar scale*, consisting of a bar broken into segments to represent distances in miles, kilometers, or some other unit of linear measurement. A graphic scale does refer to a specific unit of measurement, but has the attractive property that it remains accurate if the map is reduced or enlarged on a photocopier (assuming that the illustration of the graphic scale is reduced or enlarged, too). An *RF* fails to have this property, of course, since reducing a map diminishes the scale factor. This highlights a major difference between *RF*'s and graphic scales – a graphic scale tells us only the distance that represents a kilometer *on this map* while an *RF* also tells us how that length compares to an actual kilometer. Many atlas maps include both a graphic scale and an *RF*.

1.1. A case study. The importance of scale to the mapping process and the need for accuracy in determining and representing scale on maps cannot be stressed too strongly. As a case in point, during the summer of 1994, a colleague and practicing cartographer, Dr. Elaine Bosowski, was called upon to testify in Pennsylvania Federal Court as an expert witness in a case involving housing rights being defined by distances on maps. Her research into the case led to an interesting discovery. Maps designed to determine the distances between neighboring houses had been *altered in size* to fit a report, but the scales of the maps had not been recalculated! Indeed, the graphic scale had not been copied along with its accompanying map. As a result, a law establishing standards for distances between houses was based on a map with incorrect scale representation. As a cartographer, Dr. Bosowski felt that this alone should have been enough to get the case thrown out of court. It took hours to explain to the judge, the attorneys, and the court what had taken her only moments to discover.

2. Mathematics of scale factors

Many maps claim to have a single fixed scale factor. Yet, as we shall see in Chapter 6, no flat map of the Earth can actually have a single fixed scale. So it must be that every map we use includes a variety of scale factors. In fact, not only can the scale factor be different at different points, but the scale can also be different *in every different direction* from each point! Is it possible to figure out what the scale factors *really* are? How can we compute the scale factor in a *single direction* at a *single point* on a map?

To answer these questions, let's begin with the basic principle that, if the scale factor is M, then distances on the map are related to distances on the globe by the simple equation

$$M = \frac{\text{distance on map}}{\text{distance on globe}}.$$

To adapt this to the situation in which the scale factor is different at different points and in different directions, consider that the scale factor in any given direction can't change by too much if we restrict ourselves to only very small distances on the globe. Indeed, this is the reason that many maps, especially those showing fairly small regions, can get away with claiming to have a constant scale factor ([Monmonier, 1991]). In other words, the scale factor in any given direction from a given point is *almost constant* locally. So we can turn our simple scale factor formula into the "local" formula

$$M \approx \frac{\Delta \text{dist}_{\text{map}}}{\Delta \text{dist}_{\text{globe}}},$$

where the symbol $\approx$ means "is approximately equal to" and the symbol Δdist stands for a small increment of distance (on either the map or the globe accordingly). The exact value of M is obtained by taking the limit of the quotient on the right hand side as the increment Δdist$_{\text{globe}}$ tends to 0. **We now recognize this as the process of taking a derivative.**

More specifically, at each point on the sphere there are tangent vectors pointing in every possible direction. The local scale factor in a given direction is a directional derivative in the direction of an appropriate tangent vector.

To see how this works in practice, we will focus on just two of the many directions in which one can go – east-west along the parallel through a given point and north-south along the meridian. As we shall see later, the scale factors in these directions are often the key to determining a map's attributes.

2.1. Scale factors along a parallel. To simplify the computations, we will work not from the earth directly, but from a spherical globe whose radius is taken to be 1 unit. Also, we will focus on just two types of projections – cylindrical and azimuthal.

A *cylindrical projection* with an equatorial perspective typically shows the parallels of latitude as horizontal lines and the meridians as vertical lines equally spaced along the equator. Varying the spacing between the parallels changes the appearance as well as the mathematical properties of the map.

As mentioned before, we will measure latitudes in radians, rather than the degrees we see in atlas maps, starting from $-\pi/2$ at the south pole up to 0 at the equator and $\pi/2$ at the north pole. When dealing with cylindrical projections, the equator will be represented by a horizontal line which we take as the x-axis in the plane of the map. The parallel at latitude ϕ will be shown on the map as a segment of the horizontal line with equation $y = h(\phi)$, where h is some specific function. Note that $y = h(0) = 0$ at the equator. The most basic cylindrical projection shows the equator at its correct scale; that is, the length of the equator on the map is the same as its length on the reference globe. All other parallels are shown as horizontal segments of the same length as the equator. Thus, the overall horizontal dimension of the map will be 2π units, matching the circumference of the equator on our unit globe. To show the meridians as evenly spaced along the equator, we map the meridian at longitude θ (in radians) to the vertical line with equation $x = \theta$. Figure 2 in Chapter 2 above shows an example of a cylindrical projection. For a cylindrical projection that has two standard parallels, at latitudes $\pm\phi_0$, we need only a few simple modifications in this design.

With these conventions for a basic cylindrical projection, the horizontal distance on the map between the vertical lines representing the longitudes θ and $(\theta + t)$ is simply t. At latitude ϕ on the globe, the circumference of the parallel is $2\pi\cos(\phi)$, so the portion of the parallel that lies between longitudes θ and $(\theta + t)$ is a circular arc of length $\frac{t}{2\pi}(2\pi\cos(\phi)) = t\cos(\phi)$. Thus, the scale factor along this parallel, denoted by M_p, is equal to

$$(1) \qquad M_p = \lim_{t \to 0} \frac{\Delta\text{dist}_{\text{map}}}{\Delta\text{dist}_{\text{globe}}} = \lim_{t \to 0} \frac{t}{t\cos(\phi)} = \sec(\phi).$$

Notice that this is independent of the longitude θ; that is, the scale factor of the map is constant along any one parallel. No derivative is really needed in formula (1) because the meridians on the cylindrical projection are equally spaced along each parallel just as they are on the globe. Thus, the projection stretches every segment of the parallel at latitude ϕ by the same factor. Note that the scale factor M_p is different along different parallels, however.

For an *azimuthal projection* with a polar perspective, the globe, or some portion of it, is projected onto a plane tangent to the globe at one of the poles.

To make the computation of scale factors as straightforward as possible, we adopt the following conventions for handling azimuthal projections. First, the image of the central pole will be taken as the origin in the plane of the map. The parallels will be shown on the map as concentric circles centered at the origin, with the circle corresponding to latitude ϕ having a radius of $r(\phi)$. Notice that the function $r(\phi)$ is increasing with ϕ if the south pole is at the center, but decreasing if the north pole is central. The meridians will be portrayed as radial lines or line segments emanating from the origin. To ensure that the east/west orientation of the map is correct, the image of the meridian at longitude θ will make an angle of either $-\theta$ in case the south pole is the center of the map, or θ if the north pole is central.

On the globe of radius 1, the parallel at latitude ϕ has radius $\cos(\phi)$. Its image is a circle of radius $r = r(\phi)$ on the map. Therefore, regardless of the longitude, a change of t in the longitude angle corresponds to a distance of $\Delta\text{dist}_{\text{map}} = t\,r(\phi)$ along the image of the parallel, compared to a distance of $\Delta\text{dist}_{\text{globe}} = t\cos(\phi)$ along the parallel itself on the globe. Thus, the scale factor along this parallel is given by

$$(2) \qquad M_p = \lim_{t \to 0} \frac{\Delta\text{dist}_{\text{map}}}{\Delta\text{dist}_{\text{globe}}} = \lim_{t \to 0} \frac{t\,r(\phi)}{t\cos(\phi)} = r(\phi) \cdot \sec(\phi).$$

That no limits are really needed to compute these particular scale factors is due to the fact that the meridians on these maps are evenly spaced along each parallel just as they are on the globe. Thus, every incremental segment of the parallel is stretched by the same factor. Derivatives enter the picture for real when we compute a map's scale factors along a meridian.

2.2. Scale factors along a meridian. For a cylindrical projection, recall that we represent the equator as lying along the x-axis and the parallel at latitude ϕ as the horizontal line at height $y = h(\phi)$. On a globe of radius 1, the arc of any meridian lying between latitudes ϕ and $(\phi + t)$ has length t while its image on the map has length $h(\phi + t) - h(\phi)$. That is, $\Delta\text{dist}_{\text{map}} = h(\phi + t) - h(\phi)$, while $\Delta\text{dist}_{\text{globe}} = t$. Therefore, the scale factor, which we will denote by M_m, along the meridian at a point at latitude ϕ is given by

$$(3) \qquad M_m = \lim_{t \to 0} \frac{h(\phi + t) - h(\phi)}{t} = h'(\phi),$$

the derivative of the height function for the parallels.

Similarly, for an azimuthal projection centered on the south pole, recall that $r(\phi)$ is the radius of the circle which is the image of the parallel at latitude ϕ. The arc of the meridian on the globe between latitudes ϕ and $(\phi + t)$ has length t while the image of this segment on the map has length $r(\phi + t) - r(\phi)$. Thus,

$$(4) \qquad M_m = \lim_{t \to 0} \frac{r(\phi + t) - r(\phi)}{t} = r'(\phi).$$

A simple modification of this occurs when we make an azimuthal projection centered at the north pole, rather than the south. In this case, the derivative $r'(\phi)$ is negative because the latitude angle ϕ decreases as we move away from the north pole. Therefore, the scale factor along the meridian, which is essentially a ratio of distances and must be positive, is $M_m = -r'(\phi)$.

For both types of projections, the scale factor along a meridian depends on the latitude at which the measurement is to be made. This means that the mapping process stretches different segments of the meridian by different factors. We should really use the notation $M_m(\phi)$ instead of just M_m (but we won't).

2.3. Examples. As a first example, look at a *stereographic projection* centered on the south pole with the light source projecting from the north pole. (Figure 1 in Chapter 7 shows such a map.) As we shall discuss in Chapter 7, the circle at latitude ϕ is projected onto a circle of radius

$$r(\phi) = \frac{2\sin(\pi/2 + \phi)}{1 + \cos(\pi/2 + \phi)} = 2\tan(\pi/4 + \phi/2).$$

To determine the scale factor along the parallel at ϕ, we apply formula (2) and use some tricky trigonometry. Thus,

$$M_p = r(\phi) \cdot \sec(\phi) = 2\tan(\pi/4 + \phi/2)\csc(\pi/2 + \phi) = \sec^2(\pi/4 + \phi/2).$$

Moreover, by formula (4), the scale factor along the meridian through any point with latitude ϕ is equal to

$$M_m = r'(\phi) = \frac{d}{d\phi}\left(2\tan(\pi/4 + \phi/2)\right) = \sec^2(\pi/4 + \phi/2).$$

As it turns out, the fact that $M_p = M_m$ at every point implies that the stereographic projection is conformal. Conformal maps are discussed more in Chapter 9.

Next, consider Lambert's equal-area cylindrical projection (Figure 2 of Chapter 2). The scale factor along the parallel at latitude ϕ is $M_p = \sec(\phi)$, as we computed above. The parallel at latitude ϕ is shown here as a segment of the horizontal line at height $h(\phi) = \sin(\phi)$. Thus, $M_m = h'(\phi) = \cos(\phi)$.

Notice that $M_m \cdot M_p = 1$ at every point. As we shall discuss more in Chapter 8, this special relationship between the scale factors along the meridian and along the parallel ensures that this map is area preserving. That is, Lambert's equal-area cylindrical projection portrays the areas of all regions of the earth's surface in their correct proportions.

3. Exercises

(1) Locate the scale indicators on three different maps in an atlas. Describe in words, using correct grammar and complete sentences, what the various indicators on the map tell us about the map. What information about the scale of the map is *not* provided by the indicators?

(2) For each of the following projections, compute the scale factors M_p and M_m as functions of the latitude ϕ. Then compute numerical estimates of these scale factors for $\phi = 0$, $\phi = \pi/6$, $\phi = \pi/4$, and $\phi = \pi/2$. All projections are based on a globe of radius 1.

   (a) An *orthographic projection* of the northern hemisphere. This is an azimuthal projection with $r(\phi) = \cos(\phi)$.

   (b) A *gnomonic projection* of the northern hemisphere. This is an azimuthal projection with $r(\phi) = \cot(\phi)$.

   (c) A *plate carrée*, also called an *equirectangular projection*, is one of the simplest and oldest known map projections. This is a cylindrical projection with $h(\phi) = \phi$ (assuming that the line $x = 0$ corresponds to longitude θ).

(3) What does it mean if the scale factor of a map is equal to 1 at a certain point in a certain direction? Explain.

(4) Suppose that, for a certain map projection, one finds that $M_p(\phi_1) = \sqrt{2}$ and $M_p(\phi_2) = 2$. What does this imply about the relative appearances on the map of regions located at latitudes ϕ_1 and ϕ_2? Explain.

(5) Suppose that, for a certain map projection, one finds that $M_p(\phi_1) = (1.5)M_m(\phi_1)$. What does this imply about the appearances on the map of regions located at latitude ϕ_1? Explain.

CHAPTER 4

Distances and shortest paths on the sphere

Essentially by definition, the distance between two points is equal to the length of the shortest path connecting them. In a flat plane, the shortest path between two points is the straight line segment connecting them and the length of this segment can be measured with a ruler. On a sphere, however, there are no straight line segments. The sphere is curved in every direction (and tunneling through the earth is not allowed). So what does the shortest path between two points look like?

1. Great circles

For every sphere there is a number R and a point O in space with the property that the sphere is the surface composed of all points in space that are at a distance R from the point O. The number R is called the *radius* of the sphere and the point O is the *center* of the sphere. If we slice through the sphere, using a plane to do the slicing, the resulting cross-section is always a circle. For instance, the equator defines a plane that passes through the center of the sphere and is itself the cross-section obtained by intersecting this plane and the sphere. By symmetry, any plane that slices through the center of the sphere will produce a cross-section of the same size as the equator. Now, the radius of the equator is the same as the radius of the sphere itself, and this is the largest circle one can have on the sphere. Let's give this concept a name.

Definition: *A circle on a sphere is called a* **great circle** *if it is obtained as the intersection of the sphere and a plane containing the center of the sphere. Every great circle has a radius equal to the radius of the sphere.*

Intuitively, the larger a circle's radius is, the 'straighter' is the circle. All circles must eventually bend around by 360° but a larger circle has a bigger circumference over which to do this. A larger circle has less 'bend-per-inch' than a smaller one. You can feel this when riding a bicycle on a circular track, for instance. On a smaller circle you have to turn the wheel sharply to keep the bicycle on the track, while on a larger circle you have to turn less sharply. *The larger circle is straighter.* Quantitatively, the curvature of a circle is equal to the reciprocal of its radius, so, again, a larger circle is less curved.

The largest circles on a sphere are the great circles, so it stands to reason that the straightest and, hence, shortest route between two points on a sphere is to follow an *arc of a great circle*. To find the great circle connecting any two given points on a sphere, we first find the plane determined by these two points and the center of the sphere. The cross-section we obtain when this plane slices through the sphere is the great circle connecting the points we started with. The shorter of the two great circle arcs between the two points is the shortest route from one to the other. *Thus, great circles on the sphere play the role of straight lines in the plane,* with arcs of great circles taking the place of line segments.

It is important to note that meridians are great circles (or, more accurately, halves of great circles) but that parallels of latitude are not, except for the equator. Thus, if we wish to travel between two points at the same latitude, the shortest path between them is *not* the latitude circle. This runs counter to our visual intuition. For if two points are at the same latitude, then they are directly east or west of each other. Our intuition tells us to travel due east or west to get from one to the other but we now see that this is not the shortest way! In general, following a constant compass bearing (E, W, SE, NW, etc.) is navigationally convenient but is not the shortest route between two points. The exceptions are when the points are both on the equator or when they have the same longitude.

Example: New York City, in the United States, and Naples (Napoli), in Italy, lie very close to the parallel at 40.65° N. The distance between them, when measured along this parallel, is about 4623.5 miles or 7433.9 kilometers. However, by the formula (6) derived later in this chapter, the distance between these cities along a great circle arc is only about 4404.5 miles or 7081.6 kilometers.

Example: The cities of Quito, Ecuador, and Entebbe, Uganda, both lie on the equator. Quito is at 78° 30′ west longitude while Entebbe is at 32° 28′ east longitude. The difference in their longitudes is thus 110° 58′ ≈ 111°. Therefore, the distance between them is approximately $(111/360) \cdot (2\pi R) = 7677$ miles, where $R = 3963$ miles is the approximate equatorial radius of the earth.

Example: The cities of Bombay, India, and Islamabad, Pakistan, are on the same meridian (73° E) separated by about 15 degrees of latitude. The approximate distance between them is thus $(15/360) \cdot (2\pi) \cdot 3950 \approx 1034$ miles. Here we use the approximate polar radius of the earth, $R = 3950$ miles, in the calculation because the meridian connecting Bombay and Islamabad is an arc of a great circle passing through the poles.

Example: When travelling at sea, the principal unit of distance is the *nautical mile*. One nautical mile corresponds to one minute, or $1/60$ of a degree, of longitude along the equator. The equator is a great circle with a radius of about 3963 statute, or land, miles. Thus,

$$1 \text{ nautical mile } = \frac{(1/60)}{360} \cdot 2\pi \cdot 3963 \approx 1.1528 \text{ statute miles } = 6086.8 \text{ feet.}$$

2. The sphere *vs.* Euclid

We have just said that, in the geometry of a sphere, great circles play the rôle that is played by straight lines in Euclidean geometry, the geometry of a plane. But, as we will now see, the difference in how great circles and straight lines behave reveals a big difference between the geometry of a sphere and the plane geometry of Euclid.

Start by considering two arbitrary great circles on a sphere. Each of these is determined by a plane containing the center of the sphere. Thus, these two planes intersect each other in a line that contains the center of the sphere. This line will puncture the sphere in two points that are *antipodal* to one another; that is, these points are half-way around the world from one another, like the north and south poles. These two points are on the sphere and on both planes, so they are on both great circles. This shows that *every pair of great circles intersects in a pair of antipodal points.*

In a flat plane, the realm of Euclid's geometry, two straight lines are either parallel and, so, don't intersect at all, or they intersect in a single point. In contrast, we have just seen that on a sphere there are *no* parallel "lines" (great circles) at all! Euclid's Fifth Postulate says that, given a line L and a point P not on it, there is exactly one line through P that is parallel to L. If we try to rewrite this for the sphere, we get this: *Given a great circle L on the sphere and a point P on the sphere but not on L, then every great circle containing P intersects L (in two points).* Indeed, one way to look at the geometry of a sphere is as the set of Theorems that arise from Euclid's first four Postulates (with great circles in place of lines) together with this alternative Fifth Postulate! We will not pursue this point of view any further here.

3. A proof using vector calculus

Our intuition tells us that great circle arcs provide the shortest paths between points on a sphere. To prove that this is indeed so, we must consider two arbitrarily chosen points on a sphere and show that the great circle route connecting them is shorter than *every other* possible path. This seems a rather tall order. With a few simplifications and the help of a bit of vector calculus, however, we can manage.

First, let's assume that the sphere we are working with has radius 1 unit (of whatever units you like) and that the center of the sphere is the origin of the coordinate system. Second, by the symmetry of the sphere, let's assume that the north pole is one of the two points we wish to connect. The other point, whatever it is, can be located using latitude and longitude, so let's say that its latitude is ϕ_1, measured in radians from the equator, and that its longitude is θ_1, measured from the prime meridian. In this context, the latitude of the north pole is $\pi/2$ and its longitude is not really defined.

Now, the great circle arc from the north pole to the other point will follow the meridian on which the other point lies, namely longitude θ_1. The arc will travel an angle of $\pi/2 - \phi_1$ along this meridian, so the length of the arc will be the quantity $\pi/2 - \phi_1$ (of whatever the units are). Remember that we are assuming that the sphere has radius 1.

We must now consider some other arbitrarily chosen path connecting these same two points and show that its length is greater than (or equal to) $\pi/2 - \phi_1$ units. To do this, we can imagine ourselves travelling along the sphere from the north pole to our destination. The latitude and longitude of our location will change as we travel. In other words, we can think of our path as being determined by two functions, $\phi(t)$ and $\theta(t)$, that tell us our latitude and longitude at each time t.

To compute the length of our path, we need to integrate our speed function with respect to time over the interval of time during which we are travelling. Let's tackle our speed first.

At time t our latitude is $\phi(t)$ and our longitude is $\theta(t)$. In three-dimensional rectangular coordinates, this means that our *position vector* at time t is given by

$$\mathbf{r}(t) = \langle \cos\phi(t) \cos\theta(t), \, \cos\phi(t) \sin\theta(t), \, \sin\phi(t) \rangle.$$

Our *velocity vector* at time t is the coordinate-by-coordinate derivative of the position vector, namely,

$$\mathbf{r}'(t) = \langle -\sin\phi(t)\frac{d\phi}{dt}\cos\theta(t) - \cos\phi(t)\sin\theta(t)\frac{d\theta}{dt},$$
$$-\sin\phi(t)\frac{d\phi}{dt}\sin\theta(t) + \cos\phi(t)\cos\theta(t)\frac{d\theta}{dt}, \quad \cos\phi(t)\frac{d\phi}{dt}\rangle.$$

Our *speed* at time t is just the magnitude of our velocity vector. This is the square root of the sum of the squares of the coordinates of the velocity vector, which simplifies to

$$\|\mathbf{r}'(t)\| = \sqrt{\left(\frac{d\phi}{dt}\right)^2 + \cos^2\phi(t)\left(\frac{d\theta}{dt}\right)^2}.$$

For the final step, let's assume that our trip from the north pole begins at time $t = 0$ and concludes at time t_1. Thus, $\phi(t_1) = \phi_1$, $\theta(t_1) = \theta_1$, and $\phi(0) = \pi/2$. (The value of $\theta(0)$ isn't really defined, but it won't matter.) The distance we have to travel is equal to

$$L = \int_{t=0}^{t_1} \|\mathbf{r}'(t)\| \, dt.$$

Observe, however, that

$$\|\mathbf{r}'(t)\| \geq \sqrt{\left(\frac{d\phi}{dt}\right)^2} = \left|\frac{d\phi}{dt}\right| \geq -\frac{d\phi}{dt}.$$

It follows that

$$L \geq \int_{t=0}^{t_1} -\frac{d\phi}{dt} \, dt = \phi(0) - \phi(t_1) = \pi/2 - \phi_1.$$

The right-hand side of this inequality is the length of the great circle arc connecting our two points. Thus, we conclude that the great circle route is indeed the shortest possible!

4. Computing distances

We have seen above that we can use differences in longitude to compute the great circle distance between two points on the earth's equator. Differences in latitude can be used to find the distance between places that lie on the same or opposite meridians. With the help of some vector geometry, we can compute the great circle distance between any two points on the earth. Here's how.

Let's assume for simplicity's sake that we are working on a sphere of radius 1 unit whose center is the origin of the coordinate system. Consider two points P and Q with latitudes ϕ_1 and ϕ_2 and longitudes θ_1 and θ_2, respectively, measured in radians. Converting to three-dimensional Cartesian coordinates, we find that P and Q have rectangular coordinates

$$P = (\cos\phi_1\cos\theta_1, \cos\phi_1\sin\theta_1, \sin\phi_1) \text{ and}$$
$$Q = (\cos\phi_2\cos\theta_2, \cos\phi_2\sin\theta_2, \sin\phi_2).$$

The center of the sphere is also the center of the great circle that connects P and Q. Therefore, the vectors $\overrightarrow{OP}$ and $\overrightarrow{OQ}$ from the origin to P and to Q, respectively, are radii of the great circle. The angle between these two vectors (actually, the smaller of the two angles) tells us the portion of the great circle that must be travelled to go from one point to the other. From vector geometry, we know that the *cosine* of

FIGURE 1. Great circle route from London to Beijing

the angle between the two vectors is given by the dot product of the vectors divided by the product of the vectors' lengths. The length of each vector is just the radius of the sphere, which is 1 in this case. The value of the dot product is

$$\overrightarrow{OP} \bullet \overrightarrow{OQ}$$
$$= \cos\phi_1 \cos\theta_1 \cos\phi_2 \cos\theta_2 + \cos\phi_1 \sin\theta_1 \cos\phi_2 \sin\theta_2 + \sin\phi_1 \sin\phi_2$$
$$= \cos\phi_1 \cos\phi_2 (\cos\theta_1 \cos\theta_2 + \sin\theta_1 \sin\theta_2) + \sin\phi_1 \sin\phi_2$$
$$= \cos\phi_1 \cos\phi_2 \cos(\theta_1 - \theta_2) + \sin\phi_1 \sin\phi_2.$$

The angle between the vectors is then

$$\text{angle} = \arccos(\overrightarrow{OP} \bullet \overrightarrow{OQ})$$
(5)
$$= \arccos(\cos\phi_1 \cos\phi_2 \cos(\theta_1 - \theta_2) + \sin\phi_1 \sin\phi_2).$$

On a sphere of radius 1, the distance between the points is the same as the angle between the vectors. In general, multiply the angle by the radius of the sphere to get the distance. So, on the earth, with a mean radius R equal to 3958.76 miles or 6373.6 kilometers, we get that the distance from the point A, with longitude θ_1 and latitude ϕ_1, to the point B, with longitude θ_2 and latitude ϕ_2, along a great circle arc is given by

(6) $$\text{distance} = R\arccos(\cos\phi_1 \cos\phi_2 \cos(\theta_1 - \theta_2) + \sin\phi_1 \sin\phi_2).$$

Example: To compute the great circle distance from London, England, to Beijing, China, we take $\phi_1 = 51.5\pi/180$ and $\theta_1 = 0$ for the latitude and longitude of London measured in radians, and $\phi_2 = 2\pi/9$ and $\theta_2 = 116.35\pi/180$ for the coordinates of Beijing. By the formulas above, the angle between London and Beijing is about 1.275 radians and the distance, using the formulas just developed, is approximately 8126 kilometers or 5047 miles. The great circle route is illustrated in Figure 1.

5. Exercises

(1) (a) What is a *great circle* on a sphere?
    (b) Given any two points on a sphere, is there always a great circle that connects them? Explain.
    (c) State at least three facts, other than the definition, about great circles.
    (d) How would you convince a friend that a great circle route represents the shortest path between two points on a sphere?

(2) The cities of Macapa (in Brazil) and Pontianak (in Borneo), and the mountain Mt. Kenya (also called Kirinyaga) all lie on the equator. Use an atlas to find the longitude (to the nearest half degree) of each of these places

and then determine the distances between them. (*Hint:* The great circle connecting any two of these places is the equator itself. The longitudinal differences tell you how much of the equator lies between them.)

(3) How far is Philadelphia (USA) from the north pole? from the south pole? from the equator? from Phnom Penh (Cambodia)? (*Hint:* The meridians of Philadelphia and of Phnom Penh form a great circle.)

(4) The cities of Philadelphia, USA, and Ankara, Turkey, essentially lie on the same parallel of latitude (40° North). What is the distance between these cities travelling along the parallel? What is the distance travelling along a great circle route?

(5) On a globe of radius 1, let A be the point on the equator ($\phi_1 = 0$) at longitude $\theta_1 = \pi/4$. Let B be the point on the prime meridian ($\theta_2 = 0$) at latitude $\phi_2 = \pi/4$. **Guess** the distance between A and B along a great circle arc. Then **compute** the distance between A and B using formula (6).

(6) Every great circle is a curve in three-dimensional space and, therefore, can be parametrized. Suppose that θ_1 is the angle between two arbitrary points P and Q, viewed as vectors, on a globe of radius 1. Find a parametrization of the great circle connecting P and Q. (There are various equivalent ways of doing this. One method is to find a unit vector, say R, that lies in the same plane as P and Q and is perpendicular to P. Then, using P and R in place of $\langle 1, 0 \rangle$ and $\langle 0, 1 \rangle$, adapt the usual parametrization of the unit circle in the xy-plane.)

CHAPTER 5

Angles, triangles, and area on a sphere

1. Measuring angles on a sphere

When working in a plane, we talk about the angle between two intersecting lines. On a sphere, as we have seen, great circles take the place of lines in determining shortest routes. Moreover, as we have seen, two great circles will always intersect each other at two antipodal points. So we ought to be able to talk about the angle between any two great circles. In fact, since every great circle is obtained as the intersection of the sphere with a plane that contains the sphere's center, we *define the angle between any two great circles to be the angle between their corresponding planes.* Thus, the angle between two great circles is never greater than 90°. Observe that every meridian intersects the equator at an angle of 90°. Figure 1 illustrates the angle between two planes.

To measure the angle between intersecting *arcs* of great circles only a slight modification is needed. Assume we have two great circle arcs, each less than a semi-circle and intersecting at a common endpoint. In this case, imagine continuing the arcs until they meet again at the antipodal point to the original intersection. That is, extend the arcs into semi-circles. Now measure the angle between the two *half-planes* determined by these semi-circles and the line connecting their points of intersection, as shown in Figure 1. This angle could be as large as 180°. For instance, the meridians at 30° E and 100° W intersect at an angle of 130°.

A map projection of the globe with the property that the projected images of any two great circles (or intersecting arcs of great circles) intersect at an angle equal to that between the two great circles themselves is said to be *conformal*. Loosely, we say that a conformal map "preserves angles". The *Mercator map* and the *stereographic projection*, both discussed later, are examples of conformal maps.

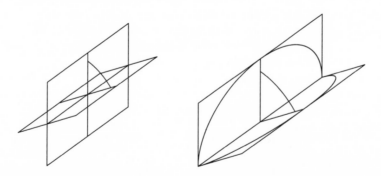

FIGURE 1. The angle between two intersecting planes and that between two great circle arcs.

2. Spherical triangles

Just as a triangle in the plane is formed by three mutually intersecting straight line segments, so a **spherical triangle** is formed when arcs of three different great circles meet in pairs. For a planar triangle, the sum of the three interior angles is always π radians but *for a spherical triangle it turns out that the sum of the interior angles can be any number of radians between π and 3π (exclusive).* To get a feel for why this is so, first note that each angle in a spherical triangle, being the angle between two great circle arcs, must be less than π radians. So the sum of the three angles must be less than 3π. Two points on the equator and a third point near the equator will define a spherical triangle that almost fills up an entire hemisphere. In that case, the sum of the angles will be just under 3π. At the other extreme, consider a very tiny triangle formed by arcs of two close meridians intersecting at the north pole and a great circle arc that intersects them near the pole. Being so small, this triangle is *almost* flat, so its angles will add up to a number close to π. Soon we shall see that this number is actually slightly greater than π. We can see how all of this works by comparing the *area* of a spherical triangle to the *sum of its angles.*

As an experiment, take a spherical globe and tie three lengths of string into loops that fit the globe snugly at the equator. These loops will be equal in size to a great circle on the globe. Practice using these loops to construct a variety of spherical triangles on the globe. For each triangle, record the sum of the angles and the area of the triangle. The area might be measured as a fraction of the total surface area $4\pi R^2$ of the globe. The following table shows just a few of the many possibilities.

spherical triangles	
angle sum (in radians)	surface area
near π	near 0
$7\pi/6$	$\pi R^2/6$
$3\pi/2$	$\pi R^2/2$
2π	πR^2
near 3π	near $2\pi R^2$

To determine the relationship between angle sum and area for a spherical triangle on a sphere of radius R, we could begin by plotting the data shown in the table in a Cartesian coordinate system, with the angle sum measured along the x-axis, say, and the area of the triangle measured along the y-axis. Then try to find a continuous curve that passes through our data points, or at least comes very close to them, and whose equation we can determine. This equation will suggest the relation between angle sum and area that we are seeking.

In this case, a careful plotting of the data from the table shows that the points all lie on a straight line. We can use the first two points from the table to determine the value, m, of the slope of the line. This yields

$$m = \frac{(\pi R^2/6) - 0}{7\pi/6 - \pi} = R^2.$$

Now use the data point $(\pi, 0)$ to get the equation of the line in point-slope form. Namely,

(7) $$y = R^2(x - \pi).$$

In this context, remember that the variable x represents the angle sum in radians of the spherical triangle and y denotes the triangle's surface area. So, by the equation (7), the relation between these two quantities is

(8)   spherical triangle area $= R^2$ (sum of the angles $- \pi$).

If the angles are measured in degrees, then equation (8) becomes

$$\text{triangle area} = \left(\frac{\pi R^2}{180}\right) \cdot (\text{sum of the angles in degrees} - 180).$$

The quantity (sum of the angles $- \pi$) is often called the *angle excess* of the spherical triangle. Our formula (8) then says that the area of a spherical triangle is proportional to the angle excess of the triangle. For instance, a triangle with one corner at the north pole and the other two lying on the equator with $1°$ of longitude separating them would have an angle excess of $(\pi/2 + \pi/2 + \pi/180) - \pi = \pi/180$. The area of the triangle would then be $A = (\pi R^2/180) \approx 273,524$ square miles, taking $R = 3958.76$ miles as the radius of the earth. This is more than the area of the state of Texas in the United States. It is also clear from the formula that the sum of the angles of a spherical triangle must exceed π. Otherwise, the triangle would have no area!

This is totally different than the situation in the plane where all triangles have the same angle sum. Two spherical triangles having the same three angles must have equal areas, while two planar triangles having the same three angles are necessarily *similar* but need not have the same area. In fact, two spherical triangles having the same three angles must have the same side lengths as well; that is, similar spherical triangles are congruent! See [Cotter, 1966] and [McCleary, 1994] for more on spherical geometry and trigonometry.

3. There are no ideal maps

We still have much to discuss about maps, but we can probably agree that for a flat map of the earth to be considered ideal or perfectly accurate all distances measured along great circle routes would have to be rescaled by the same factor (so the map would really just be a scale model of the earth), great circle routes would have to be depicted as straight line segments (so distance measurements could be made with a ruler on the map), and angles would have to be preserved (so that shapes would be depicted accurately). We now have all the tools we need to show that no flat map of a sphere, or even of a tiny portion of a sphere, can show all great circle routes between points as straight line segments *and* preserve all angles. To see this, consider a spherical triangle on the globe with corners at the points A, B, and C. The edges of the spherical triangle are arcs of great circles so, if our map shows shortest routes as straight lines, then the image of the spherical triangle must be a usual planar triangle on the map, with corners at the images of A, B, and C. But the angles of the image triangle will add up to $180°$ while we have just seen that the angles of a spherical triangle must *exceed* $180°$. So our map cannot preserve angles.

In the next chapter, we will prove that no flat map of the globe can rescale all distances measured along great circle routes by the same factor.

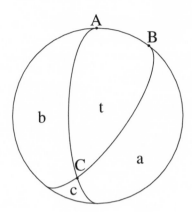

FIGURE 2. The spherical triangle with vertices at A, B and C has
area t. The areas a, b, c, and t together fill up a hemisphere. Also,
each of the areas a, b, and c forms a lune when combined with t.

4. A geometric proof of the spherical triangle area formula

For a geometric proof of the formula (8), the key is to think in terms of the
regions on the sphere formed by pairs of the great circles that make up our triangle.
These regions are called *lunes* for their resemblance to crescent moons.

Definition: A portion of the sphere between two intersecting great circles and the
antipodal points where they cross is called a *lune* of the sphere.

If a lune is formed by two great circles intersecting at an angle of α radians,
then the area of the lune is equal to $\alpha/(2\pi)$ times the total surface area of the
sphere ($4\pi R^2$ for a sphere of radius R). Thus,

$$(9) \qquad \text{area of lune} = \frac{\alpha}{2\pi} \cdot 4\pi R^2 = 2\alpha R^2,$$

where α is the angle, measured in radians, between the great circles defining the
lune.

Suppose that we now have a spherical triangle with corners at A, B, and C.
To make things easier to picture, rotate the sphere so that the edge $\overset{\frown}{AB}$ lies on
the horizon with A at the North Pole and with C in the hemisphere facing us.
Extending the arcs $\overset{\frown}{AC}$ and $\overset{\frown}{BC}$, we divide the hemisphere into four parts, as
depicted in Figure 2, with areas labelled as a, b, c, and t. The area of the triangle
is t. When we extend the arcs $\overset{\frown}{AB}$ and $\overset{\frown}{AC}$ into semicircles, we generate a lune
whose area (see equation (9)) is

$$(10) \qquad a + t = 2\,(\angle BAC)\,R^2,$$

where R is the radius of the sphere. Similarly, extending the arcs $\overset{\frown}{BA}$ and $\overset{\frown}{BC}$
generates a lune whose area is

$$(11) \qquad b + t = 2\,(\angle ABC)\,R^2.$$

A third lune is formed when the arcs $\overset{\frown}{CA}$ and $\overset{\frown}{CB}$ are extended until they intersect
at the antipode to C. This lune consists of the region t together with a copy of
region c on the back side of the sphere. The area of this lune, therefore, is

$$(12) \qquad c + t = 2\,(\angle BCA)\,R^2.$$

Adding up these three areas, we get

(13) $$a + b + c + 3t = 2R^2 \text{ (sum of the angles)}.$$

But regions a, b, c, and t form a hemisphere, so we also have

(14) $$a + b + c + t = 2\pi R^2.$$

Subtracting equation (14) from the previous equation (13) gives us

$$2t = 2R^2 \text{(sum of the angles)} - 2\pi R^2$$

so that

(15) $$t = R^2 \text{ (sum of the angles} - \pi).$$

This proves the validity of the formula we found experimentally.

5. Exercises

(1) Find the areas of the following spherical triangles: the triangle with corners at the North Pole, the city of Macapa (in Brazil), and the mountain Mt. Kenya (also called Kirinyaga); the triangle with corners at the South Pole, Mt. Kenya, and the city of Pontianak (in Borneo). Which triangle is larger? (*Hint:* All meridians intersect the equator at right angles. Each of these triangles has a corner at a pole. The angle there is the angle between the two meridians.)

(2) Find a point on the equator so that the spherical triangle defined by the North Pole, Mt. Kenya, and this other point has an angle sum of about 185°. What is the area of this triangle? (*Hint:* The hint from the previous problem applies here as well.)

(3) Find three points on the globe that define a spherical triangle whose angles add up to very nearly 540°. (*Hint:* The triangle must fill up nearly half the globe.)

(4) The **spherical Pythagorean theorem** (see [McCleary, 1994], page 5) asserts the following. *On a sphere of radius R, let $\triangle ABC$ be a spherical triangle with vertices at A, B, and C and with a right angle at the vertex C. Let a, b, and c be the lengths of the sides opposite the vertices A, B, and C, respectively. Then*

$$\cos(\frac{c}{R}) = \cos(\frac{a}{R}) \cdot \cos(\frac{b}{R}).$$

Now, on a globe of radius 1, let the points A and B be as given in exercise 5 in Chapter 4 and let C be the point where the equator and prime meridian intersect. Use the spherical Pythagorean theorem to compute the great circle distance between A and B. Compare your answer to what you got in the earlier exercise.

CHAPTER 6

Curvature of surfaces

In the previous two chapters, we learned how to measure distances along a sphere and discovered the relationship between the area and angle sum for a spherical triangle. In the hands of two of the greatest mathematicians of all time, these geometric concepts became the basis for sweeping statements about what sorts of maps of the world are possible.

1. No ideal maps, revisited

We have seen in Chapter 5 that a flat map of even a small portion of a spherical globe cannot simultaneously preserve all angles and show all shortest routes between points as straight line segments. A similar question which baffled cartographers for many years was this: Is it possible to construct a flat map of the globe, or some portion of it, which has a fixed scale? In other words, is there a flat map for which the distance between any two points on the map, measured along a straight line, is always the same multiple of the distance between the two corresponding points on the globe, measured along a great circle arc?

In symbols, if we let M denote the scale factor, then we are asking for a flat map of the globe for which the equation

$$(16) \qquad \frac{\text{distance between two points on map}}{\text{distance between same two points on globe}} = M$$

holds for every pair of points on the globe and their corresponding images on the map. Such a map would show all arcs of great circles on the sphere as straight line segments on the map *and* would have the property that the distances along these arcs were all scaled by the same factor. The map would in essence be free of distortion.

To try to construct such a map, think about peeling an orange and laying the peels out perfectly flat. It seems that we will have to tear the peels which will result in the distances across the tears being differently scaled than other distances. Even if we were to try to flatten out only a small piece of orange peel, it would seem that we would have to stretch the edges of the peel in order to avoid tearing it. But perhaps we are simply not being clever enough. After all, just because it *seems* that we can't construct such a map doesn't mean it is not possible! Indeed, over the years various attempts by cartographers to solve this problem resulted in some ingenious, if flawed, maps. Finally, in 1775, Leonhard Euler (1707–1783), the leading mathematician of his day and one of the most important mathematical figures of all time, presented to the St. Petersburg Academy of Sciences a paper entitled *On Representations of a Spherical Surface on the Plane* in which he proved conclusively that such a map could not exist.

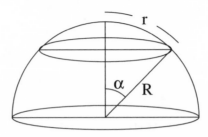

FIGURE 1. Circumference of the parallel is less than $2\pi r$ on the sphere

To see why an ideal flat map of even a portion of a sphere can't exist, we begin by using the north pole as our reference point. (By symmetry, the result would be the same no matter what point we chose.) Next, choose a small distance r. A circle is formed by the set of all those points on the sphere that lie at a distance of r away from the pole. In fact, it is a parallel of latitude since travelling a distance r from the north pole simply amounts to travelling a distance of r along any meridian. The circumference of this latitude circle is less than $2\pi r$. (Its exact circumference is $2\pi R\sin(r/R)$ which, you can check, really is less than $2\pi r$.)

Now suppose we had a map of the globe, or at least the part around the north pole, on a flat piece of paper and suppose the map scaled *all* distances by the same factor M. All the points that are at a distance of r from the north pole on the globe would get mapped to points that are at a distance of Mr from the image of the north pole on the map. So, taken together, the images of these points would form a usual Euclidean circle of radius Mr on our map. The circumference of this circle would therefore be $2\pi Mr$. But this Euclidean circle is the image on the map of our latitude circle back on the globe. Since the circumference of the latitude circle was *less than* $2\pi r$, and since all distances are being scaled by the factor M, it follows that the circumference of its image *must be less than* $2\pi Mr$. So we're stuck. The same circle can't have a circumference that is both equal to $2\pi Mr$ and less than $2\pi Mr$ at the same time. We are forced by this contradiction to conclude that **no flat map of even a portion of a sphere can have a single fixed scale.**

2. Gaussian curvature

The obstruction to flattening an orange without tearing it and never stretching it by different amounts in different places might seem quite obvious – an orange is *curved* while flat things are not. But a cone-shaped party hat is curved, too, and it *can* be flattened without wrinkling, tearing or stretching it. Just slit the hat from the open end up to the point and roll it out flat! What could be simpler? Objections: Isn't a slit really a tear? And doesn't the slit distort distances between points on opposite sides of the slit? Yes, but if we consider only a portion of the cone that's not near the slit, then we at least get a perfect flat map of that portion. Even that much is not possible for a sphere.

Let's look at another curved surface – an inflated inner tube from a tire, called a *torus*. Can we somehow flatten it without distorting distances? Suppose we try cutting the inner tube cross-wise and straightening it out into a long tube, a cylinder, open at both ends. The problem is that the circumference of the *hole* in the inner tube, where the wheel would fit, is smaller than the outer circumference, where the tire would touch the road. So when we straighten it into a cylinder, we have to stretch the inner part of the tube more than the outer part. This means that different scale factors are applied on different parts of the tube. An alternative would be to compress the outer part of the tube together, but this would create a similar problem with the scale factors. It seems that we can't really flatten an inner tube any better than we can flatten an orange. **Apparently, not all curved surfaces are curved alike.**

All of this raises the question of what we really mean by *curvature*. As with many mathematics problems, there is more than one point of view that can be taken, often with different consequences. For us the most relevant point of view is that adopted by another of the all-time great mathematicians, Carl Friedrich Gauss (1777-1855). Beginning in 1818, Gauss undertook to direct a geodetic survey of the German Kingdom of Hanover, a project that took nearly two decades to complete, though Gauss was involved for only about half of that time. Two important works grew out of this work: *Bestimmung des Breitenunterscheids zwischen den Sternwarten von Göttingen und Altona durch Beobachtungen am Ramsdenschen Zenithsektor* (1828) and *Untersuchungen über Gegenstände der Höheren Geodäsie* I and II (1843, 1846). The latter elaborates in detail the transformations necessary in order to use ordinary spherical trigonometry on an ellipsoid for the purposes of geodesy. Gauss included formulas and tables that he compiled himself to solve geodetic problems mechanically. His technique was used by geodesists throughout the nineteenth century. The *Bestimmung ...* concerns the determination of the difference in latitude between the astonomical observatories in Altona and Göttingen and features the skillful use of the method of least squares.

Gauss's work on the geodetic survey almost certainly inspired his work on differential geometry and conformal mapping, and, though perhaps not so directly, rekindled his interest in non-Euclidean geometry. In an 1822 paper submitted to the Danish Academy of Science as a prize essay, Gauss presents general conditions for a map between arbitrary areas to be conformal. Where Lambert had solved this problem for maps of the sphere to the plane, Gauss's method applies generally, including, specifically, to conformal mappings of an ellipsoid onto a sphere. This brings us to the 1827 paper *Disquisitiones generales circa superficies curvas* in which Gauss thought to ask not only how the shape of the earth might affect the geodetic measurements taken during the survey and how those measurements are interpreted, but the reverse question of what the measurements themselves would imply about the shape of the earth. The answer to this reverse question would be useful, for instance, if we lived on a cloud covered planet and could not get clues about the shape of our planet from lunar eclipses and the like.

What Gauss discovered was that one could, in fact, determine the shape of the earth, or of any surface for that matter, just by looking at certain geodetic measurements. In particular, Gauss was interested in the sum of the angles of a triangle whose sides were all segments of shortest paths on the surface, like arcs of great circles on a sphere. Gauss noticed that on many curved surfaces, including spheres, the sum of the angles would be more or less than 180°. For angle sums in excess of 180°, Gauss described the surface as having 'positive' curvature, while 'negative' curvature corresponded to angle sums of less than 180°. He introduced the concept of what we now call the *Gaussian curvature* of a surface at a given point, and, in the famous *theorema egregium* (literally, "the outstanding theorem"),

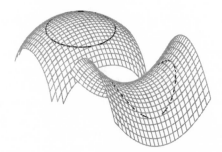

FIGURE 2. Perimeters are less than $2\pi r$ on a sphere, but greater than $2\pi r$ on a saddle.

proved that two areas can be mapped isometrically onto each other (that is, the map has a single fixed scale factor of 1) only if the Gaussian curvatures are identical at corresponding points of the two areas.

3. Quantifying curvature

In the proof given above that no flat map of a sphere can have a single scale factor, we saw that a circle on the sphere of great circle radius r centered at the north pole had a circumference of less than $2\pi r$. This idea was further developed by the French mathematicians Joseph Bertrand and Victor Puiseux who presented a quantitative version of Gaussian curvature in terms of 'circles' on the surface being studied. Begin by selecting a reference point, called P, on the surface whose curvature is to be measured. Now pick a small radius r, say, and mark on the surface all those points for which the shortest distance to P *along the surface* (that is, the path must not go into the air or tunnel beneath the surface) is equal to r. This set of marked points is the equivalent to a 'circle' on the surface centered at P. Now measure the 'circumference' of this 'circle'. If the surface is a flat plane, then we are dealing with a circle in the usual Euclidean sense and its circumference is $2\pi r$. But if the surface is a sphere, then the circumference will be less than $2\pi r$. On a saddle-shaped surface, like a saddle for a horse or the inner ring around the hole of a doughnut, the 'circumference' turns out to be *greater than* $2\pi r$. Roughly speaking, the discrepancy between the circumference measured on the surface and the value $2\pi r$ serves as a measure of the Gaussian curvature of the surface at the point P. The Gaussian curvature at a point is *positive* if the measured circumference is less than $2\pi r$ and *negative* if the measurement is greater than $2\pi r$. Thus, a sphere has positive Gaussian curvature and a saddle shape has negative Gaussian curvature. A plane and a cone both have Gaussian curvature 0; that is, they are flat. The greater the discrepancy between $2\pi r$ and the measured circumference, the greater is the amount of Gaussian curvature.

Notice that this way of measuring curvature is really a local phenomenon. The measurements are taken at the reference point P. Using a different reference point might well result in different measurements. For instance, on a torus (shaped like an inner tube or doughnut), the surface is saddle-shaped at points on the inner ring but more ball-shaped at points on the outer ring.

To be precise, the Gaussian curvature at the point P is obtained as the limiting value of the quantity

$$(17) \qquad \frac{3}{\pi} \frac{2\pi r - C(r)}{r^3},$$

where $C(r)$ is the measured 'circumference' of the 'circle' on the surface with radius r centered at P, and where the limiting value is taken as the value of r tends to 0. To get a reasonable approximation for the curvature at P, evaluate (17) for a sufficiently small choice of r.

For a sphere of radius R, the formula (17) yields a Guassian curvature of $1/R^2$ at every point. Thus, a larger sphere has a smaller Gaussian curvature – larger spheres are flatter than smaller ones. One can even think of a plane as being a sphere with an infinitely large radius!

For an inner tube with inner radius R_i and outer radius R_o, the formula gives a Gaussian curvature of $\frac{-2}{R_i(R_o-R_i)}$ at points along the hole on the inner ring of the tube and $\frac{+2}{R_o(R_o-R_i)}$ at points along the outer ring of the tube. In between, there are points where the Gaussian curvature is 0. No doubt, you can guess where they are located. On a cylinder or a cone, the Gaussian curvature is 0 at every point. These surfaces can be slit open and rolled out flat with no distortions, except for the slit itself. In other words, if we lived on a cylinder- or cone-shaped planet, then we *would* be able to make distortion-free flat maps of our world!

By providing a set of formulas that can be applied to any surface to determine the curvature at any point on the surface, Gauss laid the basis for the branch of modern mathematics known as *differential geometry*. Given a new vision by Riemann in the latter part of the nineteenth century and brilliantly applied and expanded by Einstein in the first half of the twentieth, differential geometry remains a strong mathematical discipline today. Among the interesting problems confronting geometers and physicists is that of determining the curvature of the space-time universe.

3.1. Curvature of a sphere. Let's apply formula (17) to compute the Gaussian curvature of a sphere of radius R. The circumference, $C(r)$, of the circle on the globe consisting of all points that lie at a great circle distance of r away from the north pole is given by $C(r) = 2\pi R \sin(r/R)$. We can now use the Taylor series for $\sin(x)$ to write $C(r)$ as

$$\begin{aligned} C(r) = 2\pi R \sin(r/R) \ &= \ 2\pi R \left(\frac{r}{R} - \frac{r^3}{R^3 \cdot 3!} + \frac{r^5}{R^5 \cdot 5!} - \cdots \right) \\ &= \ 2\pi r - \frac{\pi r^3}{3R^2} + \frac{\pi r^5}{60 R^4} - \cdots . \end{aligned}$$

Thus, returning to formula (17), we get that

$$(18) \qquad \lim_{r \to 0} \left(\frac{3}{\pi} \right) \frac{2\pi r - C(r)}{r^3} = \frac{1}{R^2}.$$

This is the numerical quantity which is called the *Gaussian curvature* of the sphere of radius R. Although Gaussian curvature is only a local measurement in general, the sphere has the same curvature at every point because of its symmetry.

4. Exercises

(1) At the beginning of this chapter, it is claimed that, if a flat map of a sphere had a single fixed scale factor, then the map would show all arcs of great circles on the sphere as straight line segments on the map. Here is an outline of a proof of this claim. Try to fill in the details.

   (a) Let M denote the fixed scale factor of the proposed map. Let A and B be two arbitrary points on the globe, and let r denote the distance between A and B along a great circle arc.

   (b) Let Γ denote the path on the map that is the image of the great circle arc connecting A and B. How long is Γ?

   (c) Let $\mathcal{L}$ denote the straight line segment connecting the images of A and B on the map, and let d be the length of $\mathcal{L}$. What is the length of the *pre-image* of $\mathcal{L}$ back on the sphere?

   (d) Compare the pre-image of $\mathcal{L}$ with the great circle arc connecting A and B. Then compare Γ with the straight line segment connecting the images of A and B. Conclude that $d = Mr$.

   (e) Conclude from part (d) that Γ is a straight line segment.

CHAPTER 7

Classical projections

1. Classical azimuthal projections

Recall from Chapter 2 that an *azimuthal* or *planar* projection is a map in which the earth's surface is conceptually projected onto a plane tangent to the earth at a single point. This tangent point becomes the central point on the map. With either the north or south pole as the central point, the meridians will appear on the map as straight lines radiating out from the pole. Since the meridians are arcs of great circles, we see that *the shortest path from any point on the globe* (or at least the portion of the globe being mapped) *to the central pole is shown as a straight line on the map*. Of course, any point you like can be substituted for the pole at the center, though the geometric formulas needed in the construction become more complicated. In any case, all great circles through the central point will be shown as straight lines on the map. Thus, *azimuthal maps always show shortest routes to the central point as straight lines*.

There are four azimuthal projections that were evidently known to astronomers of ancient Egypt and Greece. Three of these are projections in the literal sense and all four are still in use today.

1.1. The gnomonic map. While all azimuthal maps show shortest routes to the central point as straight lines, the *gnomonic projection*, in which the projecting light is at the center of the sphere, does much more.

Given any two points P and Q on the sphere, there is a unique great circle containing them. This great circle is realized as the intersection of the sphere with the plane determined by P, Q, and the center of the sphere O. The light source for the gnomonic projection is at the center of the sphere, so it lies in the plane that determines the great circle connecting P and Q. It follows that the image of this great circle under the gnomonic projection is the intersection of two planes – the plane that defines the great circle and the plane defined by the paper that the map is drawn on. This intersection is just a straight line on the paper. Because P and Q are on the great circle, their projections must lie on this line. Therefore, *the gnomonic projection has the property that the projection of the great circle containing any two points P and Q on the globe is the straight line connecting the projections of P and Q. That is, shortest routes on the sphere (arcs of great circles) are projected to shortest routes (straight line segments) on the (flat) map.* This holds for *all* shortest routes, not only those through the central point.

As best we can determine, the gnomonic projection was known to Thales of Miletus, a Greek merchant of the sixth century B.C. and regarded as one of the originators of the study of abstract geometry. It is worth noting that the Golden Age of Greece, in which Thales would have been an early participant, came at a time of intensified trade and great geographical discovery during which the Bronze Age gave over to the Age of Iron.

To construct a gnomonic projection, place a globe of radius R on a flat piece of paper with the south pole at the bottom and with a projecting light source at the center of the globe. Points on or above the equator won't project onto the paper, so our map will only show the southern hemisphere. We can arrange our map's coordinate axes so that the prime meridian is projected onto the positive x-axis, in which case the image of the meridian at longitude θ makes an angle of θ, measured counter-clockwise, with the positive x-axis. The parallels, meanwhile, will be shown on the map as concentric circles of various sizes having the pole as their common center. It remains only to determine the radius of the image of each parallel. For this we need to apply our knowledge of similar triangles and a bit of basic trigonometry.

When the globe is viewed from the side, as in Figure 1 in Chapter 2 above, two similar right triangles can be seen. Each has the center of the globe as a vertex. The horizontal side of the smaller triangle is a radius of the parallel at latitude ϕ. Hence, the vertical and horizontal sides of the smaller triangle have lengths $R\cos(\pi/2 + \phi)$ and $R\sin(\pi/2 + \phi)$, respectively. (Keep in mind that the angle ϕ is technically negative in this context.) The vertical side of the larger triangle is a radius of the globe, so its length is R. The length we need to determine is that of the horizontal side of the larger triangle. This length, which is the radius of the projected image of the parallel, depends on the latitude ϕ. So let's call it $r(\phi)$. The proportionality of sides for similar triangles now gives us the equation

$$(19) \qquad \frac{r(\phi)}{R} = \frac{R\sin(\pi/2 + \phi)}{R\cos(\pi/2 + \phi)}.$$

Solving this for $r(\phi)$, we get

$$(20) \qquad r(\phi) = \frac{R^2 \sin(\pi/2 + \phi)}{R\cos(\pi/2 + \phi)} = R\tan(\pi/2 + \phi) = -R\cot(\phi).$$

This is the radius of the circle on the map to which the parallel at latitude ϕ is projected, where $-\pi/2 \leq \phi < 0$ in this context. Knowing how to place the images of the meridians and the parallels, we can now create a base grid, either on the computer or "by hand" using a protractor, ruler and compass, for a gnomonic projection of most of the southern hemisphere. The resulting map is shown in Figure 1 in Chapter 2 above.

1.2. Stereographic projection. Another classical example of an azimuthal map is the *stereographic projection*. This time the globe is projected onto a flat piece of paper by means of a light source at the point antipodal to the point of contact between the globe and the paper. For instance, if the south pole is the central point, then the light source is at the north pole. The central point is its own shadow, but the shadows move farther and farther out with no bound as we move away from the central point and toward the light source. The point at which the light is located has no projected image as its shadow would fall infinitely far out from the center. Thus, the stereographic projection maps the globe, except for the light source itself, onto the infinite plane of the paper. We can think of puncturing a tiny hole in the globe where the light is located, then opening this hole wider and wider, pulling and stretching the globe out into an infinite flat sheet. It should be obvious that this projection greatly distorts areas.

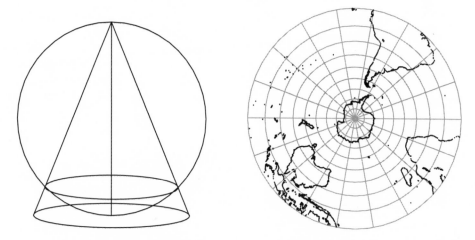

FIGURE 1. The stereographic projection: a side view and a map of the southern hemisphere.

To construct a stereographic projection centered at the south pole, we place a light at the north pole. As with the gnomonic projection, an analysis of similar triangles shows that the parallel at ϕ will be projected onto a circle centered at the south pole with radius

$$(21) \qquad r(\phi) = \frac{2R\sin(\pi/2 + \phi)}{1 + \cos(\pi/2 + \phi)} = 2R\tan(\pi/4 + \phi/2),$$

where R is the radius of the globe.

A stereographic projection of the southern hemisphere is shown in Figure 1.

1.3. Orthographic projection. In the *orthographic projection*, also handed down from antiquity, the globe is placed on a flat piece of paper and each point is projected to the nearest point on the paper as if a distant light source – "at infinity", so to speak – were being shone at right angles to the plane of the paper, with all of its light rays parallel to each other. This closely corresponds to how we would see the Earth from a distant point in space. Orthographic maps only show one hemisphere at a time to avoid superimposing the image of one hemisphere on the other.

As with all azimuthal maps, any point can be chosen as the central point depending on what portion of the globe is to be mapped. But the mathematical construction of the base grid for the map is easiest if one of the poles is used as the center. Figure 2 shows the southern hemisphere.

1.4. Azimuthal equidistant maps. As we have discussed, all azimuthal maps show great circle paths through the central point as straight lines. By adjusting the spacing between latitude parallels appropriately, we can also ensure that the *distances* along these great circle paths are correct. The resulting map is called an *azimuthal equidistant projection*. Among all azimuthal maps, this one has the special feature of showing true distances to the central point. Unlike the other azimuthal projections discussed so far, this map is a strictly mathematical construction that does not utilize a light source.

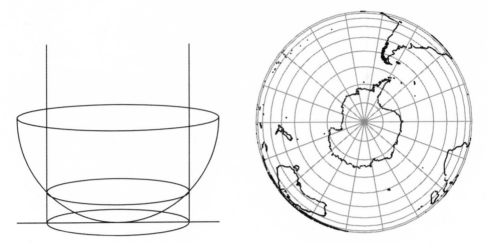

FIGURE 2. The orthographic projection: a side view and a map of the southern hemisphere.

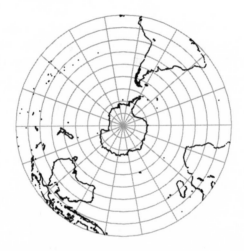

FIGURE 3. Azimuthal equidistant projection of the southern hemisphere

It is fairly straightforward to construct an azimuthal equidistant map with the south pole as its central point. The meridians, being great circles through the pole, are shown on the map as straight lines radiating out from the center. If we map the prime meridian onto the positive x-axis, then the image of the meridian at longitude θ, where $-\pi \leq \theta \leq \pi$, will make an angle of $-\theta$ with the positive x-axis. As for the parallels, all of the points on a given circle of latitude are the same distance away from the south pole. Specifically, points on the parallel at latitude ϕ, where $-\pi/2 \leq \phi \leq \pi/2$, are at a distance of $R(\pi/2 + \phi)$ from the south pole. (Here R is the radius of the reference globe.) Since the map preserves distances to the pole, the image of this parallel on the map will be a circle centered at the pole with radius equal to $R(\pi/2 + \phi)$. In particular, because the length of one degree of latitude is the same on the entire globe, the images of the parallels will be evenly spaced concentric circles on the map. Figure 3 shows the southern hemisphere.

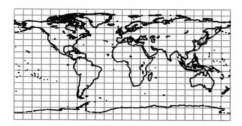

FIGURE 4. Equirectangular projection

The real value of an azimuthal equidistant map comes from the possibility of using any point we want as the central point. Then great circles through our favorite point will be shown on the map as straight lines radiating outward from that point, and the distance from any point on the globe *to our favorite point* will be shown correctly. Our favorite spot becomes the "center of the world", which explains why some have referred to this as the "egocentric map". Chapter 11 discusses how to create such a map.

2. The equirectangular projection

One of the simplest cylindrical projections is also one of the oldest, the *equirectangular projection* or *plate carée*. The meridians and parallels form a rectangular grid of identical rectangles. In the simplest construction, the images of the meridians are half as long as the image of the equator. The resulting grid is then composed of squares. Alternatively, the meridians can be depicted so that their length is correct relative to the images of the two parallels at latitudes ϕ_0 and $-\phi_0$. In any case, the overall rectangle of the map has a width-to-height ratio equal to the ratio of the length of the standard parallel to the length of any meridian. The parallels are evenly spaced along the meridians just as they are on the globe. On the globe, however, the parallels diminish in circumference toward the poles, while the equirectangular projection depicts them as horizontal segments of equal length. As is evident from Figure 4, both shapes and areas are substantially distorted in the polar regions.

Analytically, we can describe the equirectangular projection having standard parallels at latitudes $\pm\phi_0$ by mapping the point on the globe with longitude θ and latitude ϕ to the point in the Cartesian plane having x-coordinate $x = \cos(\phi_0)\theta$ and y-coordinate $y = \phi$.

3. Ptolemy's projections

During the middle part of the second century of the common era, the astronomer, geographer, and mathematician, Claudius Ptolemy of Alexandria, composed two treatises that were to have a lasting influence on Western civilization – the eight-book *Geography*, on maps and map-making, and the thirteen-book *Almagest*, on mathematics, astronomy, and the motions of the planets. Though lost to Europe during much of the Middle Ages, these works were reintroduced in the fifteenth century as the most authoritative treatments of their subject matter.

In Book 1 of *Geography*, Ptolemy introduced two original maps of what was, to him, the known world. The *oikoumene* extended from about 63° North latitude to 16°25′ South. Ptolemy's maps spanned half the globe (from about 20° west longitude to 160° east) with meridians drawn every five degrees, or every one-third of an hour change in local noon. Of course, longitude is difficult to determine accurately. In addition to providing a detailed description of the projection used for each map, Ptolemy discussed at length the importance of obtaining accurate longitude and latitude measurements for places to be depicted on the maps. Not content to rely on the accounts of travellers, he discussed the priority that must be given to astronomically determined positions and anchored his maps on well-known locations for which careful measurements had been made. He developed a variety of "error-correction" techniques for manipulating less reliable information. For instance, he devised guidelines for converting travel times at sea or on land into distance estimates and attempted to correct for the variations in wind, terrain, and direction that inevitably occured. A significant portion of *Geography* consists of lists of locations and their coordinates.

Ptolemy's first map showed the parallels as concentric circular arcs and the meridians as straight lines as far as the equator. South of the equator, Ptolemy drew the parallel at 16°25′ South to have the same length as that for 16°25′ North and angled the meridians in accordingly. This somewhat resembles a conic projection, though technically it is not. The parallels are placed so that the scale is essentially correct along the meridians. The parallels through Rhodes (about 36° North), Thule (63° North), and Syene (23°50′ North) are intended to have their correct relative lengths.

For his second map, Ptolemy retained the use of concentric, evenly spaced, circular arcs for the parallels but also drew the meridians as arcs (except for the central meridian which is a straight segment). This, he believed, was a more accurate depiction of the curved form of the earth. The parallels at Thule and Syene as well as the parallel opposite Meroë (16°25′ South) were marked off correctly for the meridians. Each set of three points was then used to generate a circular arc representing longitude. This means that the parallels are not evenly spaced along the meridians (except the central one). Interestingly, the map turns out to be nearly area-preserving.

After Columbus's voyages to the "New World", Ptolemy's projections were adapted by European mapmakers of the sixteenth century to incorporate the larger *oikoumene*.

4. Globular projections

While scientific study languished in the Europe of the Middle Ages, the classical ideas were kept alive and developed in the Islamic world. One example of a map projection developed during this period is a so-called globular projection designed for star charts by the scholar al-Biruni in about 1000 A.D. Globular projections portray one hemisphere inside a circle and, so, have the visual appeal of presenting a curved view of the earth. In al-Biruni's design, the equator and the central meridian are diameters of the outer circle, perpendicular to each other. The remaining meridians are shown as circular arcs passing through the poles and equally spaced along the equator. For the parallels, the outer circle of the map and the central meridian are marked at equal intervals. Each set of three points (two on the circle, one on the central meridian) is then joined by a circular arc. The scale is correct only along the equator and central meridian. The globular projection of one hemisphere is shown in Figure 5.

FIGURE 5. Al-Biruni's globular projection of one hemisphere.

In 1660, a Sicilian chaplain, Giovanni Battista Nicolosi, re-invented al-Biruni's projection and used it to portray the western and eastern hemispheres side-by-side. This became a standard method of showing the two hemispheres during the nineteenth century.

5. Exercises

(1) Prove that the gnomonic projection does not preserve distances. (That is, this projection does not rescale all distances by the same factor.) (*Hint:* Compare the images of two different parallels of latitude.)

(2) Determine the function $r(\phi)$ for the orthographic projection.

(3) In the construction of the azimuthal projection centered at the south pole, as described in the text, we said that, if the prime meridian is mapped onto the positive x-axis, then the image of the meridian at longitude θ, where $\pi \leq \theta \leq \pi$, will make an angle of $-\theta$ with the positive x-axis. Why $-\theta$ rather than simply θ? What if we centered the map at the north pole?

(4) At what points or along what paths is the distance scale "true" for an azimuthal equidistant projection centered at the south pole? What about for an orthographic projection centered at the south pole?

(5) Consider two equirectangular projections, one having the equator as its standard parallel (so the overall width-to-height ratio of the map is 2:1), and one having standard parallels at $\pm\pi/6$ (for a width-to-height ratio of $\sqrt{3}$:1). How will the shapes of regions compare on these two maps?

(6) In Figure 5, assume that the boundary of the map is the circle with equation $x^2 + y^2 = 1$. Then the meridian at θ is shown as a circular arc passing through the points $(0, 1)$, $(0, -1)$, and $(\theta/\pi, 0)$. Find the center, the radius, and the equation of this circle.

CHAPTER 8

Equal-area maps

If one wishes to produce a world map that displays area-based data, such as the extent of rain forests, the range of butterfly migrations, or the access of people in various regions to medical facilities, then it is often appropriate to use a base map that shows the areas of all regions of the earth's surface in their correct proportions. Such a map is called an *equal-area* or *equivalent* map by cartographers.

A few commonly used equal-area maps are the Gall-Peters equal-area cylindrical projection, Lambert's equal-area azimuthal projection, and the sinusoidal projection. The Mollweide map, which represents the whole earth on an ellipse, and Albers' equal-area conic projection, still used by the United States Geological Survey, were both presented initially in 1805.

1. Computing areas

The total surface area of a spherical globe of radius R is $4\pi R^2$, a familiar fact. More generally, the area of the portion of the sphere that lies between the equator and the parallel at ϕ radians north of the equator is equal to $2\pi R^2 \sin(\phi)$. This can be proved using calculus by taking the equation $z = \sqrt{R^2 - x^2 - y^2}$ for the upper hemisphere and computing the surface area element

$$dA = \sqrt{(\frac{\partial z}{\partial x})^2 + (\frac{\partial z}{\partial y})^2 + 1} \, dx \, dy = \frac{1}{\sqrt{R^2 - x^2 - y^2}} \, dx \, dy.$$

We then integrate the surface area element over the region of integration consisting of a ring with inner radius $r = R\cos(\phi)$ and outer radius $r = R$. Switching to cylindrical coordinates yields

$$\text{area} = \int_{\theta=0}^{2\pi} \int_{r=R\cos(\phi)}^{R} \frac{r}{\sqrt{R^2 - r^2}} \, dr \, d\theta = 2\pi R^2 \sin(\phi).$$

Notice that the area of a hemisphere, $2\pi R^2$, is obtained by taking $\phi = \pi/2$.

(To plausibly justify this formula without using calculus, start with the fact that the area of a hemisphere is $2\pi R^2$. Looking at a sphere should convince us that the area of a band bounded by the equator and the parallel at ϕ is concentrated nearer to the equator. That is, the vertical component, $\sin(\phi)$, of the latitude angle is what counts when measuring the area of the band.)

It is now easy to see that the area of the strip bounded by two parallels, at latitudes ϕ_1 and ϕ_2, is equal to $2\pi R^2[\sin(\phi_2) - \sin(\phi_1)]$ (assuming that $\phi_2 \geq \phi_1$). The portion of this strip that lies between the meridians at longitudes θ_1 and θ_2, therefore, has area

(22) $$A_{\text{block}} = R^2 \, (\theta_2 - \theta_1) \, [\sin(\phi_2) - \sin(\phi_1)],$$

assuming that $\theta_2 \geq \theta_1$.

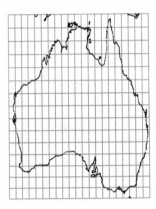

FIGURE 1. Correct areas for every block implies correct areas in general.

Finally, observe that every region on the surface of the globe can be filled up (in the limit, anyway) by some collection of (possibly infinitesimally small) blocks, each bounded by two parallels and two meridians. It follows that a given map is area preserving provided that the area of every block on the globe, bounded by any two parallels and any two meridians, is shown in its correct proportion. In fact, adding up the areas of many small blocks that fill up a larger region is exactly what an integral represents in calculus. See Figure 1.

2. Lambert's equal-area maps

Our study of equal-area maps will begin with two projections introduced in 1772 by the Swiss mathematician Johann Heinrich Lambert . For the last two decades of his fairly short life, Lambert (1728–1777) was ranked among the greatest mathematical figures of Europe. In 1761, he published the first proof that the number π is irrational. His work in geometry helped prepare the ground for the big breakthroughs in non-Euclidean geometry that were to come in the nineteenth century. He almost single-handedly pushed cartography into a new era with a 1772 treatise that, in contrast with the common approach of analyzing one map projection at a time, examined various properties that a map might exhibit and stated general conditions for their attainment. This paper also brought to bear the full power of the new analytical techniques of calculus and presented seven new map projections of which his equal-area azimuthal projection, his conformal conic, and the transverse Mercator remain important today.

2.1. Lambert's equal-area cylindrical projection. Imagine that we wrap a cylinder about a globe of radius R, just touching at the equator, and project points on the globe horizontally out to the cylinder. We then unroll the cylinder to get a flat map. In this way, the equator and all parallels will appear on the map as horizontal lines of length equal to $2\pi R$, the circumference of the equator. The meridians will be mapped as equally spaced vertical lines. To determine the vertical dimensions of the map, we observe that, because points on the globe are projected out to the cylinder horizontally, the height of a point above the equator on the map will coincide with the vertical height of the point above the plane of the equator on the globe. In particular, all points on the parallel at latitude ϕ on the globe are at a vertical height of $R\sin(\phi)$ above (or below if $\phi < 0$) the plane of the equator. Thus, on the map, this parallel is drawn at a height of $R\sin(\phi)$ above the line of the equator. Because $\sin(\pi/2) = 1$, we see that the top and bottom edges of the map are at a height of R above and below the line of the equator. The overall vertical dimension of the map is $2R$, the same as the height of the globe itself. If we now look at the portion of the map lying between the equator and the parallel at ϕ north, we see that it is a rectangular strip with dimensions $2\pi R$ by $R\sin(\phi)$ and area $2\pi R^2 \sin(\phi)$. As we saw above, this last quantity is equal to the surface area of the corresponding portion of the globe. Moreover, the meridians subdivide this strip at equal intervals just as they do on the globe, so every small block in the grid will have the correct area. Our map preserves areas!

The map we have just described is known as *Lambert's equal-area cylindrical projection* and was already pictured in Figure 2 of Chapter 2. It depicts the whole sphere on a rectangular sheet of paper with dimensions $2\pi R$ by $2R$, so the total area is $4\pi R^2$. All parallels of latitude are shown as having the same length as the equator. Thus, non-equatorial regions are stretched in width. In order to preserve their areas, these regions are compressed vertically by shrinking the space between successive parallels on the map as we move toward the poles. This is an interesting trade-off. On the globe, as one moves towards the pole, the latitude parallels are evenly spaced but get smaller in circumference. On the map, the parallels get closer together but keep the same length. It should be clear that shapes will be significantly distorted by this process, a feature that has prevented this map from seeing much use.

2.2. Lambert's equal-area azimuthal projection. To get a better sense of the broader scope and more powerful techniques of Lambert's 1772 treatise, we will use our previous work on scale factors to construct an azimuthal equal-area projection, centered at the north pole. The image of the north pole will be taken as the origin in the plane of the map. The parallels will be shown on the map as concentric circles centered at the origin, with the circle corresponding to latitude ϕ having a radius of $r(\phi)$. (Notice that the function $r(\phi)$ will actually be a decreasing function since $r(\phi)$ will increase while latitude angles decrease as we move away from the pole.) The meridian at longitude θ will be portrayed as a radial line segment emanating from the origin and making an angle of θ with the positive x-axis.

As we discussed earlier, the scale factors for this azimuthal projection are $M_p = r(\phi) \cdot \sec(\phi)$ in the direction of the parallel at latitude ϕ and $M_m = -r'(\phi)$ in the direction of the meridian through any point at latitude ϕ. This assumes that we are working from a reference globe of radius 1.

To determine the effect of scale factors on areas, consider that if we take a rectangle having adjacent sides of lengths a and b and rescale it, multiplying by M_p in the horizontal direction and M_m in the vertical direction, then the new rectangle will have adjacent sides of lengths $M_p a$ and $M_m b$. The area will have been changed by a factor of $M_p \cdot M_m$.

This works in the same way for a flat map of the sphere, provided that the map shows the parallels and meridians intersecting at right angles just as they do on the globe. This requirement is met by azimuthal projections, so it follows that **all areas will be shown in their true proportions as long as the product $M_p \cdot M_m$ is constant over the entire map.**

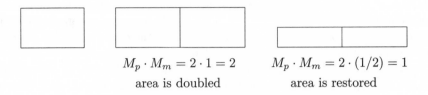

$$M_p \cdot M_m = 2 \cdot 1 = 2 \qquad\qquad M_p \cdot M_m = 2 \cdot (1/2) = 1$$

area is doubled   area is restored

FIGURE 2. Areas are affected by scale changes

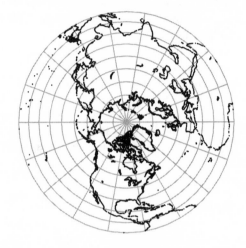

FIGURE 3. Lambert's equal-area azimuthal projection, northern hemisphere.

For instance, to create a truly area preserving map based on a globe of radius 1, we want $M_p \cdot M_m = 1$. That is, $r \sec(\phi)\,(-dr/d\phi) = 1$. This equation can be rewritten as $r\,dr = -\cos(\phi)\,d\phi$.[1] Integrating both sides yields the equation $r^2/2 = -\sin(\phi) + C$. Since $r = 0$ when $\phi = \pi/2$, it follows that $C = 1$ and, therefore, that $r^2/2 = 1 - \sin(\phi)$. We know that $r \geq 0$, so we conclude that $r = \sqrt{2}\cdot\sqrt{1 - \sin(\phi)} = 2\sin(\pi/4 - \phi/2)$. In other words, for an equal-area azimuthal projection centered at the north pole, one should map the parallel at latitude ϕ to the circle of radius $r(\phi) = 2\sin(\pi/4 - \phi/2)$ centered at the origin in the plane, and the meridian at longitude θ to the radial line segment emanating from the origin and making an angle of θ with the positive x-axis.

This projection, first presented by Johann Lambert in 1772, is known as *Lambert's azimuthal equal-area projection* and is widely used for atlas maps today (for example, see [Espenshade, et al., 1995]). Moreover, the solution $r(\phi) = 2\sin(\pi/4 - \phi/2)$ that we found is uniquely determined by the conditions that led to it. Hence, the projection we just constructed is the *only* azimuthal equal-area projection up to overall scale change.

---

[1]This is an example of a separable differentiable equation. The equation is rewritten so that each side can be integrated with respect to one of the two variables present, r and ϕ in this case.

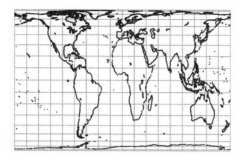

FIGURE 4. Gall-Peters equal-area map

3. The Gall-Peters map

Unlike his equal-area azimuthal projection, Lambert's cylindrical equal-area projection is rarely used because of the severe distortions of shapes, especially in the middle and upper latitudes. (See Figure 2 in Chapter 2.) One response to this problem is to redesign the projection to have two standard lines rather than only one. That is, choose a latitude ϕ_0 and think of the width of the map as corresponding to the circumference of the parallels at latitudes $\pm\phi_0$ rather than that of the equator. Then rescale everything else accordingly to preserve areas. As we shall see, such a redesigned map still exhibits distortions in the shapes of regions, but these are somewhat controlled by the fact that the map is now divided from top to bottom into three portions rather than only two.

Very briefly, to construct a modification of Lambert's equal-area cylindrical projection that has standard parallels at latitudes $\pm\phi_0$, observe that these parallels have circumference $2\pi R\cos(\phi_0)$ on the sphere of radius R. This will be the horizontal dimension of the map, with the equator appearing as a horizontal line across the middle. The meridians will be shown as evenly spaced vertical lines with the line $x = R\cos(\phi_0)\theta$ corresponding to longitude θ. As the area between the equator and the parallel at an arbitrary latitude ϕ is $2\pi R^2\sin(\phi)$, it follows that, to preserve areas, the image of the parallel must be at a height of $R\sin(\phi)\sec(\phi_0)$ above or below the equator. The total vertical dimension of the map will be $2R\sec(\phi_0)$, for a total area of $2\pi R\cos(\phi_0) \cdot 2R\sec(\phi_0) = 4\pi R^2$. One example, called the Gall-Peters projection, is shown in Figure 4. Notice that if $\phi_0 = 0$ then this construction reduces to Lambert's equal-area cylindrical projection.

For a cylindrical equal-area map having two standard parallels, the pattern of distortion is different depending on whether one is near the poles or between the two standard parallels. For instance, the Gall-Peters map, discussed more below, shows the equator to have a length of $\sqrt{2}\pi R$ when, in fact, its length is $2\pi R$. Thus, east-west distances are compressed by a factor of $1/\sqrt{2}$ at the equator. To compensate, and preserve areas, vertical distances are stretched by a factor of $\sqrt{2}$. As a result, equatorial regions appear taller and thinner in shape than they really are. The situation reverses itself near the poles, where the horizontal distances are stretched and the vertical distances compressed as with Lambert's map. Thus, polar regions appear shorter and wider than they in fact are. Distortions in shape are minimal near the standard parallels.

The choice of what parallels to use for the standard parallels is of course up to the individual cartographer. For instance, Snyder ([Snyder, 1993], page 164) cites a 1910 map of Walter Behrmann that had standard parallels at 30° North and South. Based on an analysis of distortions, Behrmann claimed that his map, used in Swedish atlases beginning in 1926, was the "best equal-area projection known for the entire world." Another variation, with standard parallels at 45° North and South, was presented in 1855 by James Gall, a Scottish clergyman. Gall's "Orthographic" projection, as he called it, became the centerpiece of a boiling controversy within the cartographic community when it was re-introduced, in 1973, by a German historian, Arno Peters. Though he apparently designed his map without knowledge of Gall's work, Peters continued to claim that it was his creation. Moreover, Peters claimed that his map at last presented the less-developed countries of the world more accurately than did the Mercator map, which he labelled as "Euro-centered". The details of this controversy provide a fascinating glimpse into how cartographers view the role of maps in society and offer compelling evidence of the power that maps have to influence peoples' thinking. (See [Harley, 2001], [Kaiser, 1993], and [Monmonier, 1995].) In our present context of studying the interactions between mathematics and mapmaking, it is worth noting that the distortions in the shape of the equatorial regions on Peters's map are of the same magnitude as the area distortions of the middle-latitude regions on Mercator's map. The question of accuracy cannot be separated from the map's purpose or use.

4. Pseudocylindrical projections

One objection that might be raised to the Gall-Peters map, or to any cylindrical projection for that matter, is that it is rectangular. That is, the earth, an essentially round object, is drawn as a very non-round rectangle. Indeed, the curved construction of Ptolemy's second projection was a main reason why he preferred it to his other maps. More recently, in the 1960's, the renowned geographer Arthur Robinson called for a move away from rectangular world maps because of the false impressions such maps can create in peoples' minds about the shapes and relative sizes of land masses, oceans, and the earth itself.

To be sure, the azimuthal maps we have considered do not show the world in a rectangle, but they have the disadvantage of having only a single central point, rather than the central parallel and meridian typical of a cylindrical projection. A class of projections known as *pseudocylindrical projections* retain something of the gridlike structure of the cylindrical projections, but try to improve the overall appearance by rounding off the edges in one way or another. We will examine the step-by-step construction of two of these pseudocylindrical projections – the Mollweide map and the sinusoidal projection. Both of these maps are area preserving and commonly used in atlases.

4.1. The Mollweide map. In 1805, Karl Brandan Mollweide (1774–1825) presented a map projection that portrays the whole world in an ellipse whose axes are in a 2:1 ratio. If one looks at a circular coin, for instance, slightly from the side rather than straight on, it presents an elliptical profile. Thus, the ellipse is a good compromise away from the rectangle and toward the circle or sphere. To add to its visual appeal, the Mollweide map incorporates the fact that, if we were to look at the earth from deep space, we would see only one hemisphere, which would appear circular to us. On this map, the hemisphere facing us as we look at the globe is depicted as a central circle, with the diameter of the circle equal to the vertical axis of the overall ellipse. The "dark side" of the earth is split in two with one piece shown on either side of the central circle. To complete the overall structure of the map, the parallels will be drawn as horizontal lines on the map (as they would appear to be if we could see them from space), while the two meridians at θ and $-\theta$ will together form an ellipse whose vertical axis coincides with the vertical axis of the overall ellipse. The meridians will be equally spaced along the equator. With this basic setup, let's turn to the problem of preserving areas.

For simplicity, we will assume we are working with a globe of radius 1. Also, the map construction given here will have as its center the point where the equator and prime meridian meet, though any point can serve as the center with the appropriate changes in the calculations. (The modifications required are minor if the central point is on the equator.)

The central circle on the Mollweide map is to represent a hemisphere, so its area must be 2π. Thus, the radius of the central circle must be $\sqrt{2}$. We will also take this as the length of the vertical semi-minor axis of the overall ellipse.

The area of an ellipse with semi-axes of lengths a and b is equal to πab. This can be proved using calculus by taking the equation $x^2/a^2 + y^2/b^2 = 1$ for the ellipse, solving for y, and integrating. By symmetry, it is enough to compute the area in the first quadrant. Thus,

$$\text{area of ellipse} = \frac{4b}{a} \int_{x=0}^{a} \sqrt{a^2 - x^2}\, dx = \pi ab.$$

For another proof, consider the matrix $T = \begin{bmatrix} a & 0 \\ 0 & b \end{bmatrix}$. A typical point on the unit circle has Cartesian coordinates $(\cos(\alpha), \sin(\alpha))$ and is transformed by T into the point $(a\cos(\alpha), b\sin(\alpha))$, which lies on our ellipse. Since transformation by the matrix T multiplies areas of regions in the plane by the factor $\det(T) = ab$, it follows that the area of our ellipse is πab as claimed.

For the Mollweide map, we have already determined that $b = \sqrt{2}$ and that the total area is to be 4π. Thus, the horizontal semi-major axis of the overall ellipse must have length $a = (4\pi)/(\sqrt{2}\pi) = 2\sqrt{2}$. That is, the overall ellipse is twice as long as it is high.

To determine the placements of the parallels, recall that the area of the portion of a sphere of radius 1 that lies between the equator and the latitude parallel at ϕ is equal to $2\pi \sin(\phi)$. Half of this, or $\pi \sin(\phi)$, will lie in the hemisphere shown in the central circle of the map. Now suppose that we place the image of the parallel for ϕ at a height of $h = h(\phi)$ above the equator on the map. (We'll assume that $\phi > 0$ here. By symmetry, $h(-\phi) = -h(\phi)$.) Let t be the angle between the equator and the radius of the central circle corresponding to the parallel, as shown in Figure 5. So $h = \sqrt{2}\sin(t)$. The area of the portion of the central circle that lies between the equator and this horizontal line (the image of the parallel) is made up of two circular sectors, each formed by an angle of t, and two right triangles, each with adjacent sides of lengths h and $\sqrt{2}\cos(t)$. Since the whole circle has area 2π, each sector has an area of

$$\text{sector area} = 2\pi \frac{t}{2\pi} = t.$$

Each of the right triangles has area

$$\text{triangle area} = \frac{1}{2} h \sqrt{2}\cos(t) = \sin(t)\cos(t) = \frac{1}{2}\sin(2t),$$

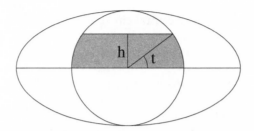

FIGURE 5. For each latitude value ϕ, the corresponding value of t
is chosen so that the shaded area in the figure is equal to $\pi\sin(\phi)$.

since $h = \sqrt{2}\sin(t)$. For our map to preserve areas, the sum of the areas of the two
sectors and the two right triangles (that is, the shaded area in Figure 5) has to be
equal to $\pi\sin(\phi)$. Thus, we have the following equation to solve for t in terms of ϕ.

$$(23) \qquad\qquad \pi\sin(\phi) = 2t + \sin(2t).$$

To get a general formula for t in terms of ϕ from this equation turns out to be
impossible, even for a computer algebra system such as Maple. Instead, we can
use the equation to numerically estimate the values of t that correspond to the
latitudes ϕ that we actually wish to show on our map. For example, substituting
$\phi = \pi/6$ into the left hand side of equation (23) and solving for t gives us $t \approx .41586$
radians (about $23.826771°$). The corresponding height for the image of the parallel
at latitude $\pi/6$ is $h = \sqrt{2}\sin(.41586) \approx .57130$. Table 1, generated using Maple,
shows the approximate values of t (in radians) and h for latitudes in increments of
$\pi/18$ radians.

$\pi\sin(\phi) = 2t + \sin(2t)$		
ϕ (radians)	t (radians)	$h = \sqrt{2}\sin(t)$
0	0	0
$\pi/18$	.13724	.19348
$2\pi/18$	.27549	.38469
$\pi/6$	.41586	.57130
$4\pi/18$	.55975	.75091
$5\pi/18$	.70911	.92088
$\pi/3$	.86699	1.07818
$7\pi/18$	1.03902	1.21892
$8\pi/18$	1.23880	1.33699
$\pi/2$	$\pi/2$	$\sqrt{2}$

TABLE 1. Placement of the parallels

For the parallels to fit exactly into the overall ellipse of the map, we need to
know how long to draw each one. To compute this, recall that the rectangular
coordinates (x, y) of each point on an ellipse with semi-axes of lengths $2\sqrt{2}$ and $\sqrt{2}$
in the horizontal and vertical directions, respectively, are related by the equation

$$(24) \qquad\qquad \frac{x^2}{8} + \frac{y^2}{2} = 1.$$

For the parallel at height h we substitute $y = h$ into this equation to get

(25)
$$\frac{x^2}{8} + \frac{h^2}{2} = 1.$$

Solving this for x in terms of h yields

(26)
$$x = \pm 2\sqrt{2 - h^2},$$

so the length of the parallel at height h is $4\sqrt{2 - h^2}$.

Having taken care of the placement of the parallels, we now turn to the placement of the meridians, which, recall, are to form ellipses when taken in pairs. We want the vertical axis of the ellipse formed by the meridians at θ and $-\theta$ to be the same as the vertical axis of the central circle. So, its length is $2\sqrt{2}$. This means that the length of the semiaxis in the vertical direction is to be $\sqrt{2}$ no matter what θ is. The length of the axis in the horizontal direction, on the other hand, will depend on θ. Our map is supposed to show correct areas, so the key to determining the right length for the horizontal axis is to look at two areas, one on the sphere and one on the map, and make sure that they are equal. First, on the map, suppose for the moment that we denote by 2α the length of the horizontal axis of the ellipse formed by the meridians at $\pm\theta$ (so α is the semi-axis in the horizontal direction). Since the vertical semi-axis is $\sqrt{2}$ and since the area of an ellipse is π times the product of the lengths of the two semiaxes, we get that the area of the ellipse on the map is

(27)
$$\text{area on map} = \pi\sqrt{2}\alpha.$$

Now look at the corresponding portion of the sphere bounded by the meridians at θ and $-\theta$. This represents an angle of 2θ radians, out of the 2π radians possible. So, the area of this region (an example of a *lune*, to repeat some earlier terminology) is the fraction θ/π of the total surface area of the sphere, taken to have radius 1. Thus,

(28)
$$\text{area on sphere} = \frac{\theta}{\pi} \cdot 4\pi = 4\theta.$$

Setting the two areas in (27) and (28) equal, we get $\pi\sqrt{2}\alpha = 4\theta$. We solve this for α in terms of θ, yielding

(29)
$$\alpha = \frac{4\theta}{\sqrt{2}\pi} = \frac{2\sqrt{2}\theta}{\pi}.$$

The ellipse we want on the map, then, has semi-axes of lengths $2\sqrt{2}\theta/\pi$ in the horizontal direction and $\sqrt{2}$ in the vertical direction. The equation of the ellipse is therefore given by

(30)
$$\frac{\pi^2 x^2}{8\theta^2} + \frac{y^2}{2} = 1.$$

Note that, if we take $\theta = \pi/2$, we get the ellipse $x^2/2 + y^2/2 = 1$ or, more simply, $x^2 + y^2 = 2$, which is the equation of the central circle just as it should be. Also, the equation of the overall ellipse for the Mollweide Map is obtained by taking $\theta = \pi$, in which case we get $x^2/8 + y^2/2 = 1$. The horizontal semi-axis of this ellipse is $\sqrt{8} = 2\sqrt{2}$ which is twice the radius of the central circle.

Figure 6 illustrates the map obtained from these equations.

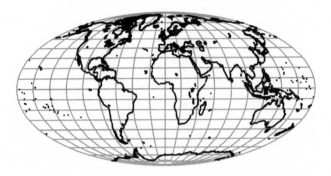

FIGURE 6. The Mollweide equal-area projection

4.2. The sinusoidal projection. A pseudocylindrical projection whose construction involves quite a bit less effort than Mollweide's map is the sinusoidal projection, which dates back at least to 1570 when it appeared in the work of Jean Cossin of Dieppe. It was used in some later editions of Mercator's atlases (1606 to 1609) and also by Nicolas Sanson d'Abbeville beginning in 1650 and by John Flamsteed (1646-1719), the first astronomer royal of England. In addition to its current appellation, a term first applied to it by d'Avezac-Macaya in 1863, the sinusoidal projection has been known variously as the Sanson-Flamsteed, the Mercator-Sanson, or the Mercator equal-area projection.

The construction of the sinusoidal map is quite simple. On a globe of radius R, the parallel at latitude ϕ has circumference $2\pi R \cos(\phi)$. This parallel is represented on the map as a horizontal line segment of width $2\pi \cos(\phi)$ placed at height ϕ above the x-axis (or below the x-axis when ϕ is negative). Thus, the lengths of the parallels are portrayed in their correct proportions. Moreover, the parallels are evenly spaced along the y-axis, just as they would be if one travelled along a meridian on the globe. The overall vertical dimension of the map is π units. The meridians are then spaced evenly along the parallels, just as they are on the globe. That is, the meridian at longitude θ is depicted as the graph of the curve $x = \theta \cos(y)$, for $-\pi/2 \le y \le \pi/2$. In particular, the outer boundary of the map consists of the curves $x = \pm\pi \cos(y)$ taken together. For any "rectangle" on the globe described by the conditions $\phi_1 \le \phi \le \phi_2$ and $\theta_1 \le \theta \le \theta_2$, the image of this rectangle on the map has area

$$\int_{y=\phi_1}^{\phi_2} (\theta_2 - \theta_1) \cos(y) \, dy = (\theta_2 - \theta_1) \left[\sin(\phi_2) - \sin(\phi_1) \right],$$

in agreement with the equation (22). Thus, areas are preserved by this projection.

5. Exercises

(1) Compute the scale factors M_p and M_m for Lambert's equal-area cylindrical projection. Show that the product $M_p \cdot M_m$ is constant throughout the map.

(2) Compute the scale factors M_p and M_m for the Gall-Peters projection. Show that the product $M_p \cdot M_m$ is constant throughout the map. (*Hint:*

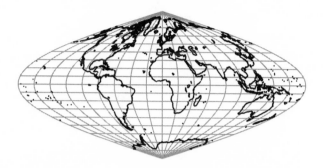

FIGURE 7. Sinusoidal projection of the world

Be careful here. The meridian at longitude θ does not map onto the line $x = \theta$ for the Gall-Peters map, so M_p is not simply $\sec(\phi)$.)

(3) Does Figure 2 apply to the Mollweide and sinusoidal maps? Why or why not?

(4) Compare the distortions in shapes and areas that you see in the Mollweide map (Figure 6) with those that you see in the sinusoidal projection (Figure 7). Don't compute anything; just compare the maps visually.

CHAPTER 9

Conformal maps

A map projection of the sphere, or of any surface for that matter, is said to be *conformal* if it preserves all angles between pairs of intersecting paths. In other words, a map is conformal if the projected images of any two intersecting paths on the surface intersect at an angle equal to that between the original paths themselves. Thus, conformal maps portray shapes accurately, at least locally. This generally comes at the expense of severe distortions in area.

To check a particular projection for conformality, it would appear from the definition that we would have to check the angles between all possible pairs of paths to verify that these angles are preserved by the projection. However, the azimuthal and cylindrical projections being considered here have the important feature that "rectangles" on the sphere, meaning regions bounded by two meridians and two parallels, are mapped to either "polar rectangles" or Cartesian rectangles in the plane. Under these circumstances, Figure 1 illustrates the fact that all angles will be preserved, and conformality achieved, if, at each point, the scale factor along the parallel, represented by the horizontal edge in the figure, is equal to the scale factor along the meridian, represented by the vertical edge in the figure.

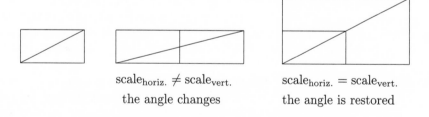

scale$_{\text{horiz.}} \neq$ scale$_{\text{vert.}}$ scale$_{\text{horiz.}} =$ scale$_{\text{vert.}}$

the angle changes   the angle is restored

FIGURE 1. How scale factors affect angles

In other words, for a given projection, if the images of the parallels and the meridians intersect at right angles as they do on the sphere and, if the scale factor M_p along the parallel is equal to the scale factor M_m along the meridian at every point, then every small rectangle on the sphere, represented by the left-most rectangle in Figure 1, will have as its image a rectangle that preserves the angle of the diagonal, represented by the right-most rectangle in Figure 1. Because every angle on the sphere is the difference between the diagonals of some pair of small rectangles on the globe, it follows that all angles will be preserved. The projection is conformal. Conversely, a conformal projection must show the parallels at right angles to the meridians and, as the diagram indicates, must satisfy the condition $M_p = M_m$ at every point.

In this chapter, we will take a closer look at two important conformal map projections of the sphere onto a plane – the stereographic projection and the Mercator projection. Within the world of mapmaking, there are, of course, other conformal maps of interest. For instance, in 1772, Lambert produced a conformal conic projection that remains widely used today by the United States Geological Survey for 7.5– and 15–minute topographic maps.

1. The stereographic projection

The *stereographic projection*, introduced earlier, is a map of the sphere that was known to the ancient Greeks at least by the time of Hipparchus of Rhodes during the second century BC. In 1695, Edmund Halley, motivated by his interest in star charts and using the newly established tools of calculus, published the first proof that this map is conformal. Recall that the stereographic projection (see Figure 1 of Chapter 7) is constructed by projecting points on the sphere onto a tangent plane from a light source located at the point on the sphere directly opposite the point of tangency. If the center of the map is the south pole $\phi = -\pi/2$, then, as we discussed before, the meridian at θ_1 projects onto the half line with polar coordinate equation $\theta = \theta_1$ while the parallel at ϕ projects onto the circle centered at the origin with radius

$$(31) \qquad r = r(\phi) = \frac{2\sin(\pi/2 + \phi)}{1 + \cos(\pi/2 + \phi)} = 2\tan(\pi/4 + \phi/2).$$

The scale factors for this projection are $M_p = r(\phi)\sec(\phi) = \sec^2(\pi/4 + \phi/2)$ and $M_m = r'(\phi) = \sec^2(\pi/4 + \phi/2)$. Thus, $M_p = M_m$ at every point. The stereographic projection is conformal.

This trait of the stereographic projection explains its usefulness for making star charts, where the observer becomes the point from which the projection emanates and the visible stars are projected onto a plane. Conformality ensures that the angles between the various constellations are the same on the chart as in the sky. Thus, the observer will know exactly where to look in the sky to locate any celestial body.

Unlike Mercator's map, the images of the meridians are not parallel to each other with the stereographic projection. Also, paths of constant compass bearing are not shown as straight lines in general. For instance, the parallels are shown as circles.

2. Mercator's map

The fact that the gnomonic projection preserves shortest routes is of tremendous value to navigators for it enables them to plot shortest routes quite easily (provided the points are not too far apart). However, to follow a great circle path generally requires continual changes in compass bearing and this is inconvenient from a navigational point of view. What Columbus did to get to the New World was to follow a parallel of latitude from the Canary Islands off the west coast of Africa until he hit land. This was not the shortest route, of course, but it had the huge advantage of actually being navigable – just make sure that the sun is at the correct angle above the horizon every day at noon. Sailing a parallel to traverse the Atlantic was common practice thereafter.

FIGURE 2. A northeast/southwest loxodrome

A better solution, had it been available, would have been to plot a route that approximated the shortest route but required only periodic changes in compass bearing. To do this, one would need, in addition to the gnomonic map, another map on which paths of constant compass bearing on the sphere were shown conveniently as straight lines. It was just such a map that the Flemish geographer Gerhard Kremer (Latin name *Gerardus Mercator*) presented in 1569 with the title *Nova et aucta orbis terrae descriptio ad usum navigantium emendate accommodata* (A new and enlarged description of the earth with corrections for use in navigation).

When following some path along the surface of the Earth, one's compass bearing at any given point on the path is represented by the angle between the direction of the path and the meridian through that particular point. These angles have traditionally been given names, such as north and south for paths that point along the meridian, or east and west for paths that are at right angles to the meridian. A path of constant compass bearing is therefore one which *makes the same angle with every meridian it crosses.*

A path of constant compass bearing on the globe is called a 'rhumb line' or 'loxodrome' and appears, in general, as a spiral converging to one of the poles. For instance, if one begins at the equator and follows the constant compass bearing 'northeast', the resulting path will spiral around the globe up to the north pole, intersecting each meridian at an angle of $\pi/4$ (45°). Figure 2 shows this path. The exceptions, of course, are headings of due east or west which keep the traveller at a constant latitude. Also, a path following a bearing of due north or due south will follow a meridian toward one of the poles and won't spiral around at all. Mercator's problem was to figure out how to show all of these spirals as straight lines on a map.

2.1. The construction. To begin the construction of Mercator's map, consider first the simplest paths of constant bearing, namely, the parallels and meridians. All parallels correspond to a constant east-west compass bearing and, so, must be shown as parallel straight lines on the map. Similarly, all meridians have a north-south orientation and, therefore, must also be shown as straight lines parallel to one another. Finally, because the east-west direction is perpendicular to the north-south direction, the images of the parallels should be perpendicular to the images of the meridians. Putting this together, it makes sense that Mercator chose for the basic form of his map a standard rectangular grid in which the equator and all parallels of latitude are shown as horizontal lines and the meridians are equally spaced vertical lines.

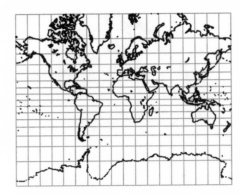

FIGURE 3. Mercator's map for latitudes $-4\pi/9 \leq \phi \leq 4\pi/9$.

The remaining problem in constructing the map is to figure out how to space the horizontal lines representing the parallels in such a way that the goal of the map is achieved. Namely, a loxodrome on the globe that makes an angle of α with every meridian it crosses should be shown on the map as a straight line that makes an angle of α with every vertical line it crosses. In order to preserve all of these angles, we clearly want to construct a conformal map. Our knowledge of scale factors will show us how to do this.

We have already seen that the scale factor along the parallel at latitude ϕ for a rectangular map is given by $M_p = \sec(\phi)$. From the condition for conformality, $M_m = M_p$, it follows that we must also have $M_m = \sec(\phi)$. If we depict this parallel as a horizontal line segment at height $y = h(\phi)$, then, as we have seen before, the scale factor along the meridian at any point on this parallel will be $M_m = h'(\phi)$. So the function $h(\phi)$ must satisfy $h'(\phi) = \sec(\phi)$. We also want to have $h(0) = 0$ so that the equator is mapped to a segment of the x-axis. Together, these conditions imply that

$$(32) \qquad h(\phi) = \int_0^\phi \sec(t)\, dt = \ln|sec(\phi) + \tan(\phi)|.$$

Notice that as we approach the north and south poles, the latitude gets close to $\pm\pi/2$ so that the scale factor $\sec(\phi)$ is tending to infinity. This explains why Mercator's map shows countries in the northern latitudes, like Greenland, to be so large compared to equatorial countries.

Taken together, the stereographic projection and Mercator's map illustrate the "inverse operation" relationship between derivatives and integrals. With the stereographic projection, the placement function $r(\phi)$ was determined by the geometric definition of the projection and we computed the derivative $r'(\phi)$ to determine the scale factor along a meridian. That this derivative was equal to the scale factor along the parallel told us that the map was conformal. For Mercator's map, the problem was reversed. We *started* by wanting to create a conformal map. In other words, we started off with a set of scale factors determined by the conformality condition. We then integrated the scale factor function to determine the placement function $h(\phi)$ for the parallels.

2.2. Integral calculus foreshadowed. The apparent ease with which we have constructed Mercator's map conceals the fact that its original formulation preceded the formal development of logarithms by almost fifty years and the invention of calculus by nearly one hundred years! Indeed, the details of how Mercator composed his remarkable map are subject to some speculation. His description of his projection, placed directly on the map over much of the portrayal of North America, says in part

> Taking all this into consideration, we have somewhat increased the degrees of latitude toward each pole, in proportion to the increase of the parallels beyond the ratio they really have to the equator.

By "increasing the degrees of latitude" is meant a stretching of the meridians, so Mercator has stated clearly here that the scale factor along the meridians must match that along the parallels. To implement this, it seems likely that he would have seen that, in order to properly place the parallels, he had to proceed in incremental steps, adjusting the (approximate) vertical scale as he went and adding the incremental height displacements that he computed. This would amount to using an approximating sum for the integral we arrived at above. Here is a look at this process.

Keep in mind that the overall form of the map is of a rectangular grid with the meridians shown as equally spaced vertical lines and the parallels shown as horizontal lines of length 2π. For the parallel at latitude ϕ, the lengthening of the parallel "beyond the ratio [it] really [has] to the equator" is $\sec(\phi)$. This is to be the vertical scale factor as well. As we move up from the equator toward the parallel at ϕ, we pass through all latitudes in between and *each latitude* has a *different* vertical scale factor.

To estimate the correct height for the image of the parallel, we will proceed from the equator to the parallel in small increments, adjusting the scale factor as we go. Let's say we decide to take small steps each of size t along the way. Our stopping points will be at

$$t, \ (2t), \ (3t), \ldots, \ (\phi - t), \ \text{and} \ \phi.$$

(Of course, we won't plot parallels for every value of ϕ, so we choose a value of t that divides evenly into all of the ϕ we care about.) The number of such small steps will be ϕ/t. (For instance, proceeding to the 20° parallel in 5° increments entails four steps with stops at 5°, 10°, 15° and 20°.) The corresponding vertical scale factors are

$$\sec(t), \ \sec(2t), \ldots, \ \sec(\phi - t), \ \text{and} \ \sec(\phi).$$

The length of the arc on the globe between successive stopping points is t in every case, so the vertical distances between the images of the stopping points on our map are (approximately)

$$t\sec(t), \ t\sec(2t), \ldots, \ t\sec(\phi - t), \ \text{and} \ t\sec(\phi).$$

To reach the image of the parallel at ϕ, we must add up all of these incremental distances. Thus, our new estimate for the height above the equator of the image of the parallel is

$$(33) \qquad t\sec(t) + t\sec(2t) + \cdots + t\sec(\phi - t) + t\sec(\phi).$$

This placement will actually be too high because the vertical scale factors *increase* as we go up from the equator and we used the scale factor at the *end* of each step to form the height estimate given in (33). We can get a height estimate

that is too low by using the scale factor at the beginning of each of the steps and
making calculations just like the ones above. In this case, the steps will start at
the latitudes

$$0, \ t, \ 2t, \ldots, \text{ and } (\phi - t)$$

with corresponding vertical scale factors of

$$1, \ \sec(t), \ \sec(2t), \ldots, \text{ and } \sec(\phi - t).$$

Now multiply each of these by the incremental distance t and add up the results to
get

$$(34) \qquad\qquad t + t\sec(t) + t\sec(2t) + \cdots + t\sec(\phi - t)$$

as a lower estimate for the placement of the parallel at latitude ϕ. The precise
location of the parallel is somewhere in between the two estimates computed in (33)
and (34), so we can get a good measure of our error by looking at their difference.
If this error is more than we are willing to accept, we can decrease the value of t
and recalculate using more and smaller increments.

Even with very small values of t (large numbers of increments) we still only get
estimates of the exact locations of the parallels. Of course, if t is small enough this
estimate will be quite satisfactory for any practical purpose. This is probably how
Mercator himself constructed his map in 1569. Nonetheless, the *exact* location of
the parallels for a map that *truly* shows loxodromes as straight lines can only be
obtained by repeating this summation procedure for smaller and smaller values of
t (and, hence, larger and larger numbers of increments) and determining a *limiting
value* of the resulting estimates. Thus, we would have to compute the limit

$$(35) \qquad\qquad \lim_{t \to 0} \sum_{k=1}^{\phi/t} t\sec(kt).$$

With the development of calculus, such a limit of approximating sums has come
to be known as an *integral*. Thus, we see a hint of the mathematics of the future
in the work of Mercator.

2.3. Navigation with Mercator's map. Equipped with both a gnomonic
map and Mercator's map, a navigator can plot a useful route as follows. First,
on the gnomonic map, draw the straight line connecting the starting and ending
points. This is the shortest possible route between the two points. Next, at regular
intervals along this route mark some reference points. Locate these same reference
points (as well as the start and finish points) on the Mercator map. Now, on the
Mercator map, connect the reference points with straight line segments. This is the
actual travel route. In this way, one follows a constant compass bearing from one
reference point to the next. The resulting route is close to the optimal (shortest)
route but requires only periodic compass bearing changes (at the reference points).
A splendid compromise!

3. Exercises

(1) Show that the gnomonic projection and the orthographic projection are
    not conformal.
(2) Are the Mollweide and sinusoidal maps conformal? Explain.

(3) Consider a hypothetical map projection for which the images of the parallels and the meridians cross at right angles. Show that projection is *both* equal-area *and* conformal if, and only if, the scale factors M_p and M_m are equal to the *same constant* throughout the map. Is such a map possible? Explain.

(4) On the stereographic projection, the images of the meridians are not parallel to each other. Use this fact to explain why the meridians are the *only* paths of constant compass bearing that are shown as straight lines by this projection.

CHAPTER 10

Analysis of map distortion

In Chapter 5, we saw that no flat map of even a small portion of the earth's surface can show all shortest routes (great circle arcs) as straight line segments and, at the same time, preserve all angles (shapes). Similarly, our analysis in Chapter 6 of the way in which the earth's surface is curved led us to conclude that no flat map of even a small portion of the earth's surface can have a single fixed scale factor that applies at all points and in all directions. In short, every flat map of even a small portion of the earth's surface must involve some distortion of shape, size, distance, or direction. It follows that, in order to understand the strengths and weaknesses of any particular map projection, and to be able to compare projections in order to select one as more appropriate for a certain use, we must tackle the essentially mathematical task of measuring and analysing distortion.

1. Length distortion

To determine how much a given map distorts lengths, the simplest approach is to directly compare lengths of paths on the globe with the lengths of their images on the map, for instance by looking at their ratio. That is,

$$(36) \qquad \text{length distortion} = \frac{\text{distance on map}}{\text{distance on globe}}.$$

You no doubt recognize this as the formula for computing a scale factor. As we have seen, the scale factor for a flat map of the globe can only really be determined locally, in one direction and at one point at a time. Since a huge effort would be required to compute all of the many various scale factors, it makes sense to single out the scale factors along the parallel and along the meridian through a given point as giving us a good indication of the distortion in length at that point. Symbolically, we will denote by M_p the scale factor along the parallel and by M_m the scale factor along the meridian. Technically, these are functions of the latitude and longitude.

For example, for the Mercator projection we computed that

$$(37) \qquad M_p = M_m = \sec(\phi),$$

where ϕ is the latitude of the point at which we are making the measurement and the radius of the globe is taken to be $R = 1$.

Similarly, for the stereographic projection centered at the south pole we have

$$(38) \qquad M_p = M_m = \frac{2}{1 - \sin(\phi)} = \sec^2(\pi/4 + \phi/2),$$

where ϕ again represents the latitude. Notice that both sets of scale factors are independent of the longitude of the point.

We can compare the length distortions of these two maps by evaluating the scale factors in equations (37) and (38) at some common points. On the equator, for instance, we get $M_p = M_m = 1$ for the Mercator map and $M_p = M_m = 2$ for the stereographic projection. Thus, length distortions at the equator are twice as great on the stereographic map as on Mercator's map. Near the south pole, we get $M_p = M_m = 1$ for the stereographic projection, but M_p and M_m both approaching ∞ for Mercator's map. At latitude $\phi = -\pi/4$, equidistant from the centers of both maps, we have scale factors of $\sqrt{2}$ (≈ 1.414) for the Mercator map and $4 - 2\sqrt{2}$ (≈ 1.172) for the stereographic projection.

As a third example consider the azimuthal equidistant projection with the south pole as its central point. This map preserves distances along the meridians since these are great circles through the pole, so that $M_m = 1$ along any meridian. At the same time, the parallel at latitude ϕ has radius $R\cos(\phi)$ on the globe but is projected to a circle of radius $R(\pi/2 + \phi)$ on the map. Thus, $M_p = (\pi/2 + \phi)\sec(\phi)$ at every point on that parallel. This yields $M_p = 1$ at the south pole, $M_p = \pi/2 \approx 1.57$ at the equator, and M_p tending to ∞ as we near the north pole, which is really a single point but is shown as a circle at the boundary of the map.

In principle, we could compute, or at least estimate, the values of M_p and M_m for any map and, in this way, formulate a sense of how much the map distorts distances.

2. Area distortion

To assess the amount of area distortion at a given point on some map, we could take a small region around that point on the globe and compare its area with the area of its image region on the map. Again, the ratio of the two areas is a reasonable way of comparing. In other words, we can measure area distortion by computing the quantity

$$(39) \qquad \text{area distortion} = \frac{\text{area on map}}{\text{area on globe}}.$$

In general, this quantity will be different at different points and so, as with length distortion, it should really be measured only locally. In practical terms, that means that we should use only fairly small regions in making our calculations.

With this measurement of area distortion, the condition for a map projection to show all areas correctly is that the ratio in (39) be equal to 1 at every point. In reality, though, we always work with a globe which is itself a scaled-down version of the earth. So we never use *truly* equal-area maps. Instead, we want map projections that show all areas in their correct proportions; that is, all areas should be rescaled by the same factor. This is the same as saying that *the area distortion factor should have the same value at every point.*

To put these ideas into practice, remember that we already know how to compute the area of any block in the graticule on the globe. Specifically, on a sphere of radius R, consider the block bounded by two parallels, at latitudes ϕ and $\phi + d\phi$, and two meridians, at longitudes θ and $\theta + d\theta$. As in equation (22), the area of this block is equal to

$$(40) \qquad A_{\text{block}} = R^2[\sin(\phi + d\phi) - \sin(\phi)]\, d\theta.$$

We can now use the concept of linear approximation from calculus to see that

$$[\sin(\phi + d\phi) - \sin(\phi)] \approx \frac{d(\sin(\phi))}{d\phi}\, d\phi = \cos(\phi)\, d\phi.$$

Incorporating this into equation (40) yields the formula

(41) $$A_{\text{block}} \approx R^2 \cos(\phi)\, d\phi\, d\theta$$

for any block at latitude ϕ on a globe of radius R.

For example, if we take $R = 1$ and $d\phi = d\theta = \pi/36$, then formula (41) gives (approximate) areas of 0.0076 when $\phi = 0$, 0.0054 when $\phi = \pi/4$, and 0.002 for $\phi = 5\pi/12$.

To determine the area distortion of a given map, we would then compare these areas of blocks in the grid on the sphere to the areas of the images of these blocks on the map. One way cartographers do this is with a tool called a *planimeter*. For cylindrical map projections, the images of the blocks are just rectangles so we can compute their areas rather easily. On azimuthal maps, the images of the blocks are sections of rings between two circles on the map and, again, their areas are not too hard to compute once we know the radii of the circles involved. As examples, let's look at Mercator's map and the azimuthal equidistant projection centered at the south pole.

On the Mercator map, the image of a block in the grid on the sphere is a rectangle. Taking 2π to be the overall horizontal width of the map and recalling that the meridians are shown as equally spaced vertical lines, we see that the width of any rectangle formed by two meridians $d\theta$ radians apart will be simply $d\theta$. To determine the height of a rectangle, we use the formula $h(\phi) = \ln|\sec(\phi) + \tan(\phi)|$ derived in equation (32) for the parallel at latitude ϕ. The height of the rectangle bounded by the parallels at ϕ and $\phi + d\phi$, therefore, is just the difference, call it Δh, between the corresponding values of h. Again we can use the concept of linear approximation to see that $\Delta h \approx h'(\phi)\, d\phi$. Thus, the area of a small rectangle in the grid of Mercator's map is given by

(42) $$A_{\text{block}} \approx h'(\phi)\, d\phi\, d\theta = \sec(\phi)\, d\phi\, d\theta$$

for any block at latitude ϕ. The area distortion inherent in Mercator's map can be measured by looking at the ratio of the area of a rectangle on the map and the area of the block that it represents in the graticule on the globe. Thus, putting the expressions (41) and (42) together, we get that the area distortion factor at latitude ϕ is approximately

(43) $$\frac{\sec(\phi)\, d\phi\, d\theta}{R^2 \cos(\phi)\, d\phi\, d\theta} = \sec^2(\phi)/R^2.$$

For example, if we take $R = 1$, $d\phi = d\theta = \pi/36$, and $\phi = 0$, then formula (43) gives an area distortion factor of 1. That is, there is no distortion at the equator for this map. For the same values for R, $d\theta$, and $d\phi$, the area distortion factors are $\sec^2(\pi/4) = 2$, when $\phi = \pi/4$, and $\sec^2(5\pi/12) \approx 14.9$ for $\phi = 5\pi/12$.

On the azimuthal equidistant projection centered at the south pole, we must look at the "polar rectangle" bounded by two circles, with radii $r(\phi)$ and $r(\phi + d\phi)$, and two rays separated by an angle of $d\theta$. The exact area of this block is

$$\pi\left[r(\phi + d\phi)^2 - r(\phi)^2\right](d\theta/2\pi).$$

However, for small values of $d\phi$, we have

$$r(\phi + d\phi)^2 - r(\phi)^2 = [r(\phi + d\phi) - r(\phi)][r(\phi + d\phi) + r(\phi)] \approx r'(\phi)d\phi \cdot 2r(\phi),$$

where we have used linear approximation yet again. Hence, the area of a small polar rectangle is approximately $r(\phi) \cdot r'(\phi) \, d\phi \, d\theta$ (or simply $r \, dr \, d\theta$, which might look familiar if you have studied double integrals in calculus). Of course, for the azimuthal equidistant projection, the function r is given by $r(\phi) = R(\pi/2 + \phi)$ so that $r'(\phi) = R$. This means that the area of a small block in the grid of the azimuthal equidistant projection is given by

$$(44) \qquad A_{\text{block}} \approx r(\phi) \cdot r'(\phi) \, d\phi \, d\theta = R^2(\pi/2 + \phi) \, d\phi \, d\theta$$

for any block at latitude ϕ. Comparing this to the area of the same block on the globe of radius R yields an area distortion factor of

$$(45) \qquad \frac{R^2(\pi/2 + \phi) \, d\phi \, d\theta}{R^2 \cos(\phi) \, d\phi \, d\theta} = (\pi/2 + \phi) \, \sec(\phi).$$

If we again take $R = 1$, $d\phi = d\theta = \pi/36$, and $\phi = 0$, then formula (45) gives an area distortion factor of $\pi/2$. That is, the areas of regions near the equator are magnified by a factor of about 1.57 for this map. For the same values for R, $d\theta$, and $d\phi$, the area distortion factors are about 1.11 when $\phi = -\pi/4$ and 1 for $\phi = -\pi/2$. Comparatively, at latitude $-\pi/4$ (i.e., 45° S), which is equally far away from the centers of the two maps, the azimuthal equidistant projection has significantly less area distortion than the Mercator map (a factor of 1.11 compared to 2).

You may have noticed in the examples we just looked at that, in both cases, the area distortion factors we computed were equal to the product of the scale factors $M_p \cdot M_m$. In fact, this works for all of the cylindrical and azimuthal projections we have considered as well as any projection for which the images of the parallels and meridians intersect at right angles. To see why, consider that, if a rectangle with adjacent sides of lengths a and b is rescaled, multiplying by the scale factors M_p in the horizontal direction and M_m in the vertical direction, then the area of the rectangle will change by a factor of $M_p \cdot M_m$. This idea works in the same way for a flat map of the sphere, provided that the map shows the parallels and meridians intersecting at right angles. In other words, **the quantity $M_p \cdot M_m$ provides a measure of the area distortion factor at each point on the map, at least for certain special types of projections.**

In this special case, the condition for a map projection to show all areas correctly is that $M_p \cdot M_m = 1$ at every point. For a map projection to show all areas in their correct proportions, the area distortion factor, $M_p \cdot M_m$, should have the same value at every point.

Tables 1 and 2 give the area distortion factors for some of the projections we have discussed.

For the Mercator map, for instance, regions at latitudes around $\pm\pi/4$ will appear on the map to be twice as large as they actually are relative to regions of the same area near the equator, but only one-half of their size relative to regions of the same area at latitude $\pm\pi/3$. Also, if keeping area distortions under control is a more important consideration than conformality, these measurements suggest that the azimuthal equidistant projection might be a better choice than the stereographic projection or the Mercator map.

projection	area distortion factors			
	$M_p \cdot M_m$	$\phi = -\pi/3$	$\phi = -\pi/4$	$\phi = 0$
stereographic	$\sec^4(\pi/4 + \phi/2)$	1.15	1.37	4
gnomonic	$-\csc^3(\phi)$	1.54	2.83	∞
orthographic	$-\sin(\phi)$	.866	.707	0
equidistant	$(\pi/2 + \phi)\sec(\phi)$	1.047	1.11	1.57
Lambert's azi.	1	1	1	1

TABLE 1. Area distortion at latitude ϕ for selected azimuthal projections centered at the south pole

projection	area distortion factors			
	$M_p \cdot M_m$	$\phi = 0$	$\phi = \pm\pi/4$	$\phi = \pm\pi/3$
Mercator	$\sec^2(\phi)$	1	2	4
Lambert's cylind.	1	1	1	1

TABLE 2. Area distortion for cylindrical projections

3. Angle distortion

As with distortions in length and area, we could hope to measure the distortion in angle for a given map by looking at the ratio of an angle formed by two intersecting paths on the globe and the angle between the images of those paths on the map. Thus,

$$(46) \qquad \text{angle distortion} \; = \frac{\text{angle on map}}{\text{angle on globe}}.$$

The angles could be measured with a protractor, perhaps.

Again, this distortion factor is really a local measurement. Much as different directions can be scaled differently by a map projection, even if they emanate from the same point, so different angles based at a given point can be distorted differently. In the case of length distortions, we got around this problem by choosing the latitudinal and longitudinal directions as being somehow special and using the scale factors in those directions as representative of the overall picture. We could do something similar with angle distortions by choosing paths through a given point making angles of, say, $\pi/6$, $\pi/3$ and $\pi/2$ radians with the meridian through that point and measuring the angles between the images of those paths and the image of the meridian on the map.

Since conformal maps do not distort angles, a closer look at the conformal maps we know might help us to find an alternative approach to measuring angle distortions on other maps. Both Mercator's map and the stereographic projection are conformal. That is, these maps depict angles correctly and, at least locally, show the shapes of land masses accurately. Remember that what makes this work is that, for each of these maps, *the scale factor along the parallel at each point is equal to the scale factor along the meridian at the same point*. This suggests that a reasonable way of measuring distortion in angle for *any* map projection might be to compare the scale factors of the projection along the parallel and along the meridian at each point. That is, with M_p denoting the scale factor along a parallel, and M_m denoting the scale factor along a meridian, we would like to represent the

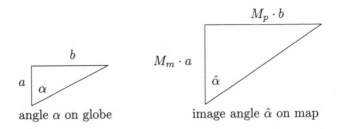

FIGURE 1. $\tan \alpha = b/a$; $\tan \hat{\alpha} = (M_p b)/(M_m a)$; $M_p/M_m = \tan \hat{\alpha}/\tan \alpha$.

angle distortion factor by the ratio M_p/M_m, where this ratio can have different values at different points. A projection will be conformal exactly when this ratio is equal to 1 at every point and the images of the parallels and meridians intersect at right angles.

The ratio M_p/M_m works as a measure of angle distortion only for maps on which the images of the parallels and meridians intersect at right angles just as they do on the globe. Fortunately, this is the case for many of the maps we have studied, including all cylindrical and azimuthal projections. It does not hold for the Mollweide map or for the sinusoidal projection, however.

Technically, the ratio M_p/M_m measures the ratio between the *tangents* of an angle on the globe and its image angle on the map. This is illustrated in Figure 1.

In the first part of the figure, the angle α represents an angle on the globe between a meridian and some other great circle. The ratio of the horizontal side of the rectangle to the vertical side is $\sin \alpha / \cos \alpha$, or $\tan \alpha$ in other words. In the second part of the figure, the angle $\hat{\alpha}$ is the projected image on the map of the original angle α. The sides of the rectangle are now $M_p \sin \alpha$ and $M_m \cos \alpha$, so we get that

$$\tan \hat{\alpha} = \frac{M_p \sin \alpha}{M_m \cos \alpha} = \frac{M_p}{M_m} \tan \alpha;$$

that is,

$$\frac{\tan \hat{\alpha}}{\tan \alpha} = \frac{M_p}{M_m}.$$

The map has to show parallels and meridians intersecting at right angles for the figure in the second part of the diagram to be a rectangle. That's why this approach only works in special cases.

projection	angle distortion factors			
	M_p/M_m	$\phi = -\pi/3$	$\phi = -\pi/4$	$\phi = 0$
stereographic	1	1	1	1
gnomonic	$-\sin(\phi)$	.866	.707	0
orthographic	$-\csc(\phi)$	1.155	1.414	∞
equidistant	$(\pi/2 + \phi)\sec(\phi)$	1.047	1.11	1.57
Lambert's azi.	$\sec^2(\pi/4 + \phi/2)$	1.072	1.172	2

TABLE 3. Angle distortion at latitude ϕ for selected azimuthal projections centered at the south pole

projection	angle distortion factors			
	M_p/M_m	$\phi = 0$	$\phi = \pm\pi/4$	$\phi = \pm\pi/3$
Mercator	1	1	1	1
Lambert's cylind.	$\sec^2(\phi)$	1	2	4

TABLE 4. Angle distortion for some cylindrical projections

Tables 3 and 4 give the angle distortion factors at various latitudes for some of the projections we have discussed. The more the distortion factor differs from 1, the greater is the angle distortion. This information can be useful to us in selecting a projection. For instance, the function $\sec^2(\phi/2)$ grows considerably more slowly than the function $\sec^2(\phi)$, making Lambert's azimuthal equal-area projection significantly more useful than his equal-area cylindrical projection. The azimuthal equidistant projection has an angle distortion factor that grows even more slowly, but lacks the desirable feature of being area preserving.

4. Tissot's Indicatrix

So far in this chapter, we have made a good start at tackling the task of measuring and analyzing map distortions. We have seen that the quantities $M_p \cdot M_m$ and M_p/M_m can be used to measure distortions of areas and angles, respectively, at least for maps that show the parallels and meridians intersecting at right angles. In this section, we take a look at another method for analyzing distortions developed in the nineteenth century by the French mathematician, Tissot (see [Snyder, 1993]). Called **Tissot's indicatrix**, this tool is today a cartographic standard for quantifying and, especially, visualizing distortions in angles and areas. Many introductory cartography textbooks include some mention of Tissot's method, but few attempt to apply it to any real map projections beyond a brief indication of how to interpret the diagrams that the indicatrix yields.

The starting point for this technique is Tissot's observation that, for any map projection, there is at each point on the sphere a pair of perpendicular directions whose images in the projection are also perpendicular. Tissot called these the *principal directions* at the given point, and measured distortion using the scale factors of the projection along these directions. It is not necessarily easy to determine what the principal directions are for a given map. However, for the cylindrical and azimuthal projections that we have considered here, the principal directions are just the parallels and the meridians. Thus, for these maps, we have *already computed* the scale factors needed to implement Tissot's program.

Visually, Tissot's indicatrix consists of a system of ellipses drawn directly on the map itself as follows. At each point, we determine the principal directions and compute the scale factors in those directions. We then draw an ellipse, centered at the given point, whose principal axes are aligned with the principal directions and have lengths equal to twice the scale factor in the corresponding direction. In practice, a representative selection of points is used and the ellipses are actually rescaled by a common factor so that they fit on the map and don't interfere with each other too much. This way, it is possible to make effective visual comparisons between different ellipses in the indicatrix.

For a cylindrical or azimuthal projection of the sphere, we draw at each point on the map an ellipse whose principal axes have lengths $2 \cdot M_p$ and $2 \cdot M_m$. The

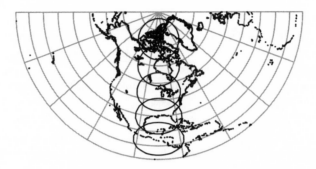

FIGURE 2. Tissot's indicatrix for Lambert's equal-area azimuthal projection. All ellipses have the same area, but are more elongated away from the pole.

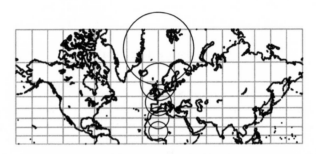

FIGURE 3. Tissot's indicatrix for Mercator's projection. All ellipses are circles, but the areas increase away from the equator.

area of such an ellipse is just $\pi M_p M_m$. Hence, the factor $M_p M_m$ represents the ratio of the area of the ellipse to the area of a circle of radius 1. The more $M_p M_m$ varies from the value 1, the more the map distorts areas.

For an equal-area projection, the condition that $M_p M_m = 1$ implies that the ellipses have the same area over the whole map. More generally, the variation in the areas of the ellipses over the range of any given map is an indicator of the area distortions inherent in the map.

At the same time, the ratio M_p/M_m represents the ratio of the lengths of the principal axes of the ellipses in the indictarix. If this ratio is equal to 1 at every point, then the projection is conformal and every ellipse is actually a circle. The more elongated an ellipse is, the more this ratio varies from the value 1 and, thus, the more angles are distorted at that point. (Again, this assumes that the projection involved shows the parallels and meridians intersecting at right angles.)

4.1. Maximum angle error. Tissot also used the principal scale factors to measure, at each point on the projection, the *maximum error* between any angle and its image angle. To see how to do this, look again at Figure 1 above. Solving for the image angle, $\hat{\alpha}$, we get $\hat{\alpha} = \arctan((M_p/M_m)\tan\alpha)$. If $M_p = M_m$, this implies that $\hat{\alpha} = \alpha$ as it should for a conformal map. If $M_p \neq M_m$, then we find the maximum error between $\hat{\alpha}$ and α by taking the derivative, with respect to α, of the error function $\hat{\alpha} - \alpha = \arctan((M_p/M_m)\tan\alpha) - \alpha$, setting the result equal to 0, and solving for α. This shows that the maximum error occurs when

$\alpha = \arctan(\sqrt{M_m/M_p})$. This is for angles measured from a meridian. An arbitrary angle is the difference between two such angles, so we have to double the error we just computed to get the overall maximum angle error. Tissot denoted the maximum angle error by 2ω.

For the projections we have considered, the scale factors depend on the latitude, so the maximum angle error at any given point will depend only on the latitude of the point. Table 5 gives a comparison of the maximum angle errors encountered at selected latitudes for various azimuthal projections. As with the angle distortion factor computed above, we see that, among the non-conformal projections considered, the azimuthal equidistant projection performs relatively well as does Lambert's equal-area azimuthal projection.

Projection	maximum angle error, 2ω		
	at $\phi = -\pi/3$	at $\phi = -\pi/4$	at $\phi = 0$
gnomonic	$-.046\pi$	$-.11\pi$	$-\pi$
orthographic	$.046\pi$	$.11\pi$	π
equidistant	$.0145\pi$	$\pi/30$	$.143\pi$
Lambert's azimuthal	$.022\pi$	$.05\pi$	$.216\pi$

TABLE 5. Maximum angle error for some non-conformal azimuthal projections centered at the south pole

5. Exercises

(1) For the azimuthal equidistant projection centered at the south pole, we have computed the scale factor along the parallel at latitude ϕ to be $M_p = (\phi + \pi/2) \cdot \sec(\phi)$. At the south pole, where $\phi = -\pi/2$, we therefore get $M_p = 0 \cdot \infty$, an indeterminate form. In fact, we should have used a limit:
$$M_p(-\pi/2) = \lim_{\phi \to -\pi/2^+} (\phi + \pi/2) \cdot \sec(\phi).$$
Show that this limit is equal to 1.

(2) Compute the (approximate) area distortion factors for the stereographic projection (centered at the south pole) for regions near the south pole, in the mid-latitudes of the southern hemisphere, and near the equator. For the same three regions, show that the area distortion factors you get for the Gall-Peters map are all (approximately) equal to 1. What does this tell you about the Gall-Peters map?

(3) In the text, it is pointed out that the angle distortion factor is a local phenomenon, possibly varying even for different angles based at the same point. In order to test this notion, take a globe and, at each of a few different points, choose paths making angles of 30, 60 and 90 degrees with the meridian through that point. Then, on a Mollweide map, measure the angles between the images of those paths and the image of the meridian on the map. Compute the angle distortion factor $\tan(\hat{\alpha})/\tan(\alpha)$ for each angle.

(4) Repeat the previous exercise for an azimuthal equidistant projection. What *should* happen if you try this exercise using the stereographic projection?

(5) Compute the ratio M_p/M_m at several different latitudes for each of (i) Lambert's cylindrical equal-area projection; (ii) the orthographic projection; (iii) the gnomonic projection.

(6) Suppose that you wish to make a conformal map of Australia. Which projection would show areas more accurately – Mercator or stereographic? Explain.

(7) Suppose that you wish to make an equal-area map of the United States of America. Which projection would show shapes (angles) more accurately – Gall-Peters or Lambert's Azimuthal projection (centered at the north pole)? Explain.

(8) (Revisit exercise 4 of Chapter 3.) Suppose that, for a certain map projection, one finds that $M_p(\phi_1) = \sqrt{2}$ and $M_p(\phi_2) = 2$. What does this imply about the appearances on the map of regions located at latitudes ϕ_1 and ϕ_2? Do you have enough information to say anything about distortions in angles or areas? Could the map possibly be area preserving? Could the map possibly be conformal? Explain.

(9) (Revisit exercise 5 of Chapter 3.) Suppose that, for a certain map projection, one finds that $M_p(\phi_1) = (1.5)M_m(\phi_1)$. What does this imply about the appearances on the map of regions located at latitude ϕ_1? Could the map possibly be area preserving? Could the map possibly be conformal? Explain. What is the angle distortion factor for this map at latitude ϕ_1? At a point at latitude ϕ_1, what would be the angle between the image of the meridian and the image of a northeast loxodrome?

(10) Does there exist a flat map of the earth for which Tissot's indicatrix consists entirely of identical circles? of identical non-circular ellipses? Explain your answers.

(11) Referring to section 4.1, verify that the maximum angle occurs when $\alpha = \arctan\left(\sqrt{M_m/M_p}\right)$. Then compute the maximum angle error, 2ω, in terms of M_p and M_m.

CHAPTER 11

Oblique perspectives

In previous chapters, our maps have always been centered at a pole, in the case of an azimuthal projection, or, for a cylindrical projection, at some point on the equator. In real life, one may prefer to have a map centered somewhere else. For instance, a pilot navigating in the northern latitudes might want a version of Mercator's map that is centered at the north pole while a researcher in the Amazon might want to use Lambert's equal-area azimuthal projection but have it centered at the equator. You might like to have an azimuthal equidistant projection based at your hometown in order to easily measure distances from there to any other location. In this chapter, we shall see how to change the center of a world map to any point we wish.

1. Changing the coordinate framework

1.1. Azimuthal projections. Let's say that we want to adapt an azimuthal projection, normally centered at the north pole, so that it will be centered instead at a point A of our own choosing. By pretending that A is the north pole, we can form in our minds an imaginary set of "meridians" – great circles emanating from A and passing through the point antipodal to A. Likewise, we can imagine an "equator" lying halfway between A and its antipode, and "parallels" subdividing the imaginary meridians. In this new framework, every point on the globe can be assigned a new pair of longitude and latitude coordinates. In order to construct our azimuthal map centered at A, we would simply apply the usual map equations to these new longitude and latitude values. This, of course, begs the question of how we can actually compute longitude and latitude of an arbitrary point on the globe relative to a new centering point A. The answer is provided by a basic construction from linear algebra using three-by-three matrices.

For simplicity, let us suppose that the point A, where our new map will be centered, is in the northern hemisphere with longitude θ_0 and latitude ϕ_0 (so $\phi_0 \geq 0$). Thus, A has Cartesian coordinates

$$A = \langle \cos(\theta_0) \cos(\phi_0),\ \sin(\theta_0) \cos(\phi_0),\ \sin(\phi_0) \rangle.$$

In our imaginary framework of meridians emanating from A, let us take A's real meridian, at θ_0, to be the new "prime meridian". (We could declare an international law to implement this, for instance!) This meridian intersects the new equator at the point B with longitude θ_0 and latitude $\phi_0 - \pi/2$. Using a trigonometric identity or two, the Cartesian coordinates of B are $\langle \cos(\theta_0) \sin(\phi_0),\ \sin(\theta_0) \sin(\phi_0),\ -\cos(\phi_0) \rangle$. With A taking the place of the north pole, the point B plays the role of the point where the prime meridian and equator intersect. In terms of Cartesian coordinates, A is substituting for the point $\langle 0, 0, 1 \rangle$ and B is taking the place of the point $\langle 1, 0, 0 \rangle$. To complete the transition to the new framework, we need a point, C

let's call it, that we can substitute for $\langle 0, 1, 0 \rangle$. This is provided by the vector cross product of A and B. Again using some trigonometric identities, we get that $C = A \times B = \langle -\sin(\theta_0), \cos(\theta_0), 0 \rangle$. To repeat, the points A, B, and C taken together form a new perpendicular three-dimensional framework that we shall use in place of the usual coordinate axes. The point A replaces the north pole ($\langle 0, 0, 1 \rangle$); the point B lies on the new 'equator' taking the place of $\langle 1, 0, 0 \rangle$; the point C lies at the intersection of the actual and new equators, perpendicular to both A and B, and plays the role of $\langle 0, 1, 0 \rangle$. We now wish to calculate the coordinates of all other points on the globe relative to this new coordinate framework we have established.

One of the central ideas of linear algebra is that a three-by-three matrix may be interpreted as a linear transformation of three-dimensional space in which the standard basis vectors $\langle 1, 0, 0 \rangle$, $\langle 0, 1, 0 \rangle$, and $\langle 0, 0, 1 \rangle$ are transformed into the vectors that comprise the columns of the matrix. In our context, we can define the matrix

$$(47) \qquad S = \begin{bmatrix} \cos(\theta_0)\sin(\phi_0) & -\sin(\theta_0) & \cos(\theta_0)\cos(\phi_0) \\ \sin(\theta_0)\sin(\phi_0) & \cos(\theta_0) & \sin(\theta_0)\cos(\phi_0) \\ -\cos(\phi_0) & 0 & \sin(\phi_0) \end{bmatrix}$$

that has the vectors B, C, and A as its columns. When viewed as a transformation of three-dimensional space, the matrix S has the effect of mapping the standard vectors $\langle 1, 0, 0 \rangle$, $\langle 0, 1, 0 \rangle$, and $\langle 0, 0, 1 \rangle$ onto the vectors B, C, and A, respectively. However, for our oblique map of the world, centered at A, we want the transformation to go the *other way*. We want A to become the point $\langle 0, 0, 1 \rangle$ where S turns $\langle 0, 0, 1 \rangle$ into A. The matrix that represents the transformation we want is therefore the *matrix inverse* of S, so called because the transformation it implements acts to reverse the effect of S. When the transformations of S and its inverse are carried out one after the other the overall effect is that every vector in three-dimensional space is back where it started. Computationally, the inverse of the matrix S is

$$(48) \qquad T = S^{-1} = \begin{bmatrix} \cos(\theta_0)\sin(\phi_0) & \sin(\theta_0)\sin(\phi_0) & -\cos(\phi_0) \\ -\sin(\theta_0) & \cos(\theta_0) & 0 \\ \cos(\theta_0)\cos(\phi_0) & \sin(\theta_0)\cos(\phi_0) & \sin(\phi_0) \end{bmatrix}.$$

We can use the matrix T just defined to compute the relative longitude and latitude, in the new framework determined by A, B, and C, of any point on the globe. For instance, the actual north pole has relative Cartesian coordinates

$$T \begin{bmatrix} 0 \\ 0 \\ 1 \end{bmatrix} = \begin{bmatrix} -\cos(\phi_0) \\ 0 \\ \sin(\phi_0) \end{bmatrix}.$$

The relative longitude and latitude for the north pole are, therefore, π and ϕ_0, respectively. (Section 6 of Chapter 1 discussed how to recover longitude and latitude from Cartesian coordinates.)

To actually create a map with a non-standard centering point, we take the usual formulas for the projection we want and apply them to the relative longitude and latitude values. In other words, we form a composition of functions where standard longitude and latitude are converted into relative coordinates which in turn are projected according to the map formulas. Figure 1 shows an azimuthal equidistant projection centered at Tokyo, Japan, while Figure 2 is an orthographic map centered on Mexico City.

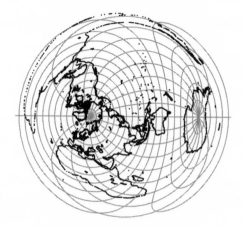

FIGURE 1. Azimuthal equidistant projection centered at Tokyo.

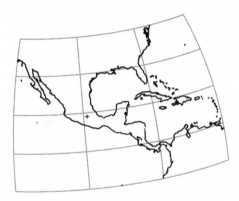

FIGURE 2. Orthographic projection centered on Mexico City.

1.2. An example. To construct the map shown in Figure 1, we first find the longitude and latitude of Tokyo. Converted to radians, these coordinates are $\theta_0 = 7\pi/9$ and $\phi_0 = \pi/5$. Tokyo is the point A. Using Cartesian coordinates and rounding off the decimals yields $A = \langle -.62, .52, .59 \rangle$. The point B has longitude $7\pi/9$ and latitude $-3\pi/10$ which in rounded-off Cartesian coordinates is $B = \langle -.45, .38, -.81 \rangle$. The point $C = A \times B$ has coordinates $C = \langle -.64, -.77, 0 \rangle$. Thus,

$$S \approx \begin{bmatrix} -.45 & -.64 & -.62 \\ .38 & -.77 & .52 \\ -.81 & 0 & .59 \end{bmatrix}.$$

The matrix that converts usual coordinates into relative coordinates, where Tokyo takes the place of the north pole, is

$$T = \begin{bmatrix} -.45 & .38 & -.81 \\ -.64 & -.77 & 0 \\ -.62 & .52 & .59 \end{bmatrix}.$$

For an arbitrary point on the sphere, with longitude θ and latitude ϕ, let's write

$$T \begin{bmatrix} \cos(\theta)\cos(\phi) \\ \sin(\theta)\cos(\phi) \\ \sin(\phi) \end{bmatrix} = \begin{bmatrix} t_1(\theta, \phi) \\ t_2(\theta, \phi) \\ t_3(\theta, \phi) \end{bmatrix}.$$

That is,

$$
\begin{aligned}
t_1(\theta, \phi) &= -.45\cos(\theta)\cos(\phi) + .38\sin(\theta)\cos(\phi) - .81\sin(\phi) , \\
t_2(\theta, \phi) &= -.64\cos(\theta)\cos(\phi) - .77\sin(\theta)\cos(\phi) , \text{ and} \\
t_3(\theta, \phi) &= -.62\cos(\theta)\cos(\phi) + .52\sin(\theta)\cos(\phi) + .59\sin(\phi).
\end{aligned}
$$

As discussed in section 6 of Chapter 1, the relative latitude is $\phi_{\text{rel}} = \arcsin(t_3(\theta, \phi))$, while the relative longitude, θ_{rel}, satisfies $\cos(\theta_{\text{rel}}) = t_1(\theta, \phi)/\sqrt{1 - t_3(\theta, \phi)^2}$ and $\sin(\theta_{\text{rel}}) = t_2(\theta, \phi)/\sqrt{1 - t_3(\theta, \phi)^2}$. Finally, recall that an azimuthal equidistant projection centered at the north pole maps the point with longitude θ and latitude ϕ to the point with two-dimensional Cartesian coordinates $x = (\pi/2 - \phi)\cos(\theta)$ and $y = (\pi/2 - \phi)\sin(\theta)$. When we substitute the expressions for *relative* longitude, θ_{rel}, and *relative* latitude, ϕ_{rel}, into the standard mapping, we get the map in the figure.

1.3. Cylindrical projections. To construct a cylindrical projection centered not where the equator and prime meridian intersect but at some other point, we can follow the same procedure used for oblique azimuthal projections with only a few modifications. Starting with the same desired center A that we used before (in the northern hemisphere), we now want to substitute A for the point $\langle 1, 0, 0 \rangle$. Travelling northward along the meridian one quarter of the way around, we reach the point $B = \langle -\cos(\theta_0)\sin(\phi_0), -\sin(\theta_0)\sin(\phi_0), \cos(\phi_0) \rangle$, with longitude $\theta_0 \pm \pi$ and latitude $\pi/2 - \phi_0$. The point B will be our substitute for the north pole $\langle 0, 0, 1 \rangle$. Finally, the point $C = B \times A = \langle -\sin(\theta_0), \cos(\theta_0), 0 \rangle$ will take the place of $\langle 0, 1, 0 \rangle$. Viewed as a transformation of three-dimensional space, the matrix

$$(49) \qquad S = \begin{bmatrix} \cos(\theta_0)\cos(\phi_0) & -\sin(\theta_0) & -\cos(\theta_0)\sin(\phi_0) \\ \sin(\theta_0)\cos(\phi_0) & \cos(\theta_0) & -\sin(\theta_0)\sin(\phi_0) \\ \sin(\phi_0) & 0 & \cos(\phi_0) \end{bmatrix},$$

that has the vectors A, C, and B as its columns, has the effect of mapping the standard vectors $\langle 1, 0, 0 \rangle$, $\langle 0, 1, 0 \rangle$, and $\langle 0, 0, 1 \rangle$ onto the vectors A, C, and B, respectively. As before, we want the transformation to go the other way, so we use the matrix inverse of S, given this time by

$$(50) \qquad T = S^{-1} = \begin{bmatrix} \cos(\theta_0)\cos(\phi_0) & \sin(\theta_0)\cos(\phi_0) & \sin(\phi_0) \\ -\sin(\theta_0) & \cos(\theta_0) & 0 \\ -\cos(\theta_0)\sin(\phi_0) & -\sin(\theta_0)\sin(\phi_0) & \cos(\phi_0) \end{bmatrix}.$$

For instance, relative to the new framework, the actual north pole has longitude 0 and latitude $\pi/2 - \phi_0$ and relative Cartesian coordinates $\langle \sin(\phi_0), 0, \cos(\phi_0) \rangle$. Notice that this is just the third column of the matrix T.

1.4. Scale factors and distortion analysis. As we have seen, oblique maps are just rotated versions of maps that have a natural polar or equatorial perspective. The key to the rotation is to compute relative longitude and latitude coordinates with respect to the new center. It follows that the pattern of scale factors and map distortions, both in area and angle, for the oblique map will be a rotation of the scale and distortion pattern for the same projection in its natural perspective. That is, the scale factors that prevail at a certain latitude for a standard projection will prevail at the same *relative latitude* for the oblique map. The distortion factors will also be the same, but must be computed using relative longitude and latitude. Thus, the distortion tables in Chapter 10 can also be used for oblique maps, but must be reinterpreted as referring to latitude and longitude relative to the new centering point of the map.

2. The transverse Mercator map

One of the most important maps presented by Lambert in his seminal 1772 treatise is the transverse Mercator projection. The term *transverse* is used to designate a map whose usual perspective has been rotated by ninety degrees. Thus, a transverse variant of an azimuthal projection would be centered on the equator while, as in the case we will examine here, a tranverse variant of a cylindrical map will be centered at one of the poles.

To construct a transverse Mercator projection centered at the north pole, say, we follow the procedure described above. The latitude of the north pole is $\phi_0 = \pi/2$ and we will take $\theta_0 = 0$ as the central longitude, though any longitude value is acceptable. The central point on the map, A, is just the north pole, or $\langle 0, 0, 1 \rangle$. Therefore, the point B is $\langle -1, 0, 0 \rangle$ and $C = B \times A = \langle 0, 1, 0 \rangle$. Thus, the matrix that we need to translate coordinates into the new framework is given by

$$T = \begin{bmatrix} 0 & 0 & 1 \\ 0 & 1 & 0 \\ -1 & 0 & 0 \end{bmatrix}.$$

The point with usual longitude and latitude θ and ϕ, respectively, has Cartesian coordinates $\langle \cos(\theta)\cos(\phi), \sin(\theta)\cos(\phi), \sin(\phi) \rangle$. Applying T to this, we get the relative Cartesian coordinates $\langle \sin(\phi), \sin(\theta)\cos(\phi), -\cos(\theta)\cos(\phi), \rangle$.

Recalling the formulas for the Mercator map, the x-coordinate is just equal to the longitude of the point being mapped. For the transverse Mercator, we will use the *relative* longitude. For the hemisphere closest to the central point, the relative longitude, $\tilde{\theta}$, is equal to $\arctan(y_{\text{rel}}/x_{\text{rel}})$, or

$$\tilde{\theta} = \arctan\left(\frac{\sin(\theta)\cos(\phi)}{\sin(\phi)} \right)$$

in our case. This is valid in the northern hemisphere, so whenever $\phi > 0$. (For the southern hemisphere, we must either add or subtract π to this value to get the correct relative longitude. To avoid this complication, we'll just plot the northern hemisphere.) The y-coordinate for Mercator's map is a function of the latitude. In our context here, the relative latitude is $\tilde{\phi} = \arcsin(z_{\text{rel}})$, or $\tilde{\phi} = \arcsin(-\cos(\theta)\cos(\phi))$. The y-coordinate is then $y = \ln\left| \sec(\tilde{\phi}) + \tan(\tilde{\phi}) \right|$. Computationally, this looks like a bit of a nightmare, but nothing that a good calculator can't manage. Figure 3 shows a transverse Mercator projection of the northern hemisphere.

In the standard Mercator map, the north and south poles are not depicted – they are sent off to infinity. For the transverse Mercator, the point $B = \langle -1, 0, 0 \rangle$ and its antipode $-B = \langle 1, 0, 0 \rangle$ get the honors. On the globe, these points lie where the prime meridian and its companion the International Date Line cross the equator. Consequently, on the map, the images of these two meridians can't cross

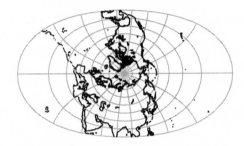

FIGURE 3. Transverse Mercator projection of the northern hemisphere, latitude $\phi \geq \pi/18$.

the image of the equator (they "meet" at infinity). At the same time, these images are straight lines. Thus, the equator and the prime meridian are shown on the transverse Mercator map as parallel straight lines.

It is clear from the picture that paths of constant compass bearing are not shown as straight lines on the transverse Mercator. Among the parallels of latitude, only the equator is a straight line on the map. The only meridians shown as straight lines are the prime meridian and the International Date Line. When we changed the centering point to the north pole and computed the relative coordinates of points on the globe, all compass bearings got changed to *relative* compass bearings. The directions "relative north" and "relative south" mean moving along a great circle route through the point B, while "relative east" and "relative west" lie perpendicular to these great circles. In other words, "relative compass bearings" are interpreted relative to treating the point B as the north pole. Thus, the transverse Mercator map actually shows all great circles through B as straight lines. Of these, only the equator is a path of constant compass bearing in the usual sense.

3. Exercises

(1) Referring to example 1.2, in which Tokyo, Japan, takes the place of the north pole, compute the relative latitudes and longitudes of the following locations: the north pole; the south pole; Greenwich, England; Tokyo, Japan.

(2) Using example 1.2 as a guide, determine the matrices S and T you would use to construct an orthographic projection centered at Mexico City. (Such a map is shown in Figure 2.) Use the matrix T to determine the relative latitude and longitude of an arbitrary point, where Mexico City is now taking the place of the north pole. Finally, compute the map projection formulas for this map. (Recall that an orthographic projection of the northern hemisphere maps the point (θ, ϕ) on the sphere to the point in the plane with Cartesian coordinates $(\cos(\phi)\cos(\theta), \cos(\phi)\sin(\theta))$.)

(3) In equations (47) and (49), notice that each matrix S has the property that its columns form an orthonormal basis for $\mathbf{R}^3$; that is, each column of S is a unit vector, and any two columns have a dot product equal to 0. Notice also that, in each case, the inverse of S is equal to the transpose of S; that is, $S^{-1} = S^t$. **Prove** the following statement about matrices. *An $n \times n$ matrix S, with real number entries, satisfies the condition $S^{-1} = S^t$ if, and*

only if, the columns of S form an orthonormal basis for $\mathbf{R}^n$. (Terminology: Such a matrix S is called a *unitary* matrix.)

CHAPTER 12

Other worlds: Maps of surfaces of revolution

The sphere is obviously of special interest to us because of its resemblance to the shape of our planet, but other surfaces of revolution can also be mapped using similar methods. This is so because it is possible to define latitude and longitude in a natural way for such surfaces. As a result, we can define maps that are analogous to our azimuthal and cylindrical maps of the sphere.

1. Surfaces of revolution

To generate a basic surface of revolution, we begin with a function f, defined and continuous on some (open or closed, finite or infinite) interval I, such that f is continuously differentiable on the interior of I and satisfies $f(x) > 0$ there. The surface S_f is then formed by revolving the curve $y = f(x)$ on the interval I about the x-axis one complete revolution.

The surface S_f can be parametrized by $(x, f(x)\cos(\theta), f(x)\sin(\theta))$ for x in I and for $-\pi \leq \theta \leq \pi$. Fixing a value of θ produces a *meridian* of the surface S_f. Thus, the meridians are obtained by intersecting the surface with any plane (in xyz-space) that contains the x-axis. Each meridian is a copy of the profile curve $y = f(x)$. Each fixed value of x in I corresponds to a *parallel* of S_f. Thus, the parallels are the intersections of the surface with planes perpendicular to the x-axis. Each parallel is a circle that is perpendicular to every meridian. Notice that this way of identifying the parallels is different than what we have been using for the sphere, where each parallel was distinguished by its latitude, measured as an angle away from the equator. In this more general context, the latitude angle is replaced by the position of the cross-section along the surface.

2. Cartographic projections of S_f

By analogy with the spherical setting, a *cylindrical projection* of the surface S_f will show the meridians as vertical lines or line segments and the parallels as horizontal line segments. Specifically, the meridian corresponding to the parameter value θ_0 will be shown as a segment of the line $x = \theta_0$. For the parallels, we first select an increasing function $h(x)$. The parallel corresponding to the parameter value x_0 will be shown as a segment of width 2π of the horizontal line $y = h(x_0)$. Thus, just as on the surface itself, the map images of the meridians are "evenly spaced" and are perpendicular to the images of the parallels. The spacing between the parallels is altered in passing from the surface to the map image. As in the case of the sphere, this spacing is what determines the map's mathematical properties.

An *azimuthal map* of S_f is constructed using polar coordinates in the plane. The meridian corresponding to the angle θ_0 on S_f will be shown as a segment of the half-line $\theta = \theta_0$ while the parallel obtained by intersecting S_f with the plane

$x = x_0$ will be shown as a circle with radius $r = r(x_0)$, where the function r can be either increasing or decreasing but must in any case be non-negative.

To get a solid footing, let us see how some by-now-familiar maps of the sphere fit into this new context. We start by generating a sphere of radius 1 as a surface of revolution. To do this, take the profile curve $f(x) = \sqrt{1 - x^2}$ for $-1 \le x \le 1$ and revolve its graph about the x-axis. Thus, the points at $x = \pm 1$ correspond to the north and south poles with the circular cross-section at $x = 0$ representing the equator.

In Mercator's map of the sphere, the parallel at latitude angle ϕ was placed as a horizontal line at height $y = h(\phi) = \ln|\sec(\phi) + \tan(\phi)|$. In the new context, though, the latitude angle ϕ is replaced by the x-coordinate at which the appropriate cross-section is made. This corresponds to $x = \sin(\phi)$. Using some right-triangle trigonometry, we see that, if $x = \sin(\phi)$, then $\sec(\phi) = 1/\sqrt{1 - x^2}$ and $\tan(\phi) = x/\sqrt{1 - x^2}$. Thus, $\sec(\phi) + \tan(\phi) = (1 + x)/\sqrt{1 - x^2} = \sqrt{(1 + x)/(1 - x)}$. On the map, this means that the parallel at latitude x will be shown as a horizontal line segment at height

$$(51) \qquad\qquad y = h(x) = \ln\left|\sqrt{\frac{1 + x}{1 - x}}\,\right| = \frac{1}{2}\ln\left|\frac{x + 1}{x - 1}\right|.$$

As we would expect, this expression for $h(x)$ tends to $\pm\infty$ as x approaches the values ± 1, which means that it is not possible to show the poles on the map. The images of the meridians are vertical lines, with the meridian at angle θ corresponding to the vertical line $x = \theta$ on the map.

For the stereographic projection, we project the sphere onto the plane $x = -1$, tangent to the south pole, from a light source at the north pole $x = 1$. As before, the meridian at angle θ gets mapped onto the half-line emanating from the origin at the same angle θ. We found earlier (formula (21)) that the parallel at latitude angle ϕ was projected onto a circle of radius $r = r(\phi) = 2\tan(\pi/4 + \phi/2)$. With $x = \sin(\phi)$ and some creative use of trigonometric identities (!), this amounts to

$$(52) \qquad\qquad r = r(x) = 2\sqrt{\frac{1 + x}{1 - x}}.$$

3. Scale factors

For an arbitrary surface of revolution, we may want to be able to analyze the distortions inherent in a particular map. Or, we might wish to construct a map that is conformal or equal-area. To accomplish these goals, we need to be able to compute scale factors, just as we did for maps of the earth.

Recall that the scale factor of a map is a ratio of lengths — the length of a path on the map divided by the length of the actual path that it represents. Moreover, this is a local phenomenon that can be measured accurately using calculus. For maps of surfaces of revolution, we will compute the scale factors along the meridian and along the parallel through any given point.

The fact that the meridian of the surface S_f at θ_0 is shown either as the vertical line $x = \theta_0$ or as the polar half-line $\theta = \theta_0$ means that each incremental segment of any given parallel is stretched by the same factor. Therefore, no derivatives are needed to compute the scale factor along a parallel. The parallel at x on S_f is a circle of circumference $2\pi f(x)$ which is stretched or compressed to a line segment of length 2π on a cylindrical projection and to a circle of circumference $2\pi r(x)$ on an azimuthal projection. Thus, the scale factor along this parallel, denoted simply by M_p, is equal to

$$(53) \qquad M_p = \frac{2\pi}{2\pi f(x)} = \frac{1}{f(x)}$$

for the cylindrical projection and

$$(54) \qquad M_p = \frac{2\pi r(x)}{2\pi f(x)} = \frac{r(x)}{f(x)}$$

for the azimuthal projection. On the sphere, for instance, $f(x) = \sqrt{1-x^2}$ and $x = \sin(\phi)$, so either $M_p = 1/f(x) = \sec(\phi)$ or $M_p = r(x)/f(x) = r \cdot \sec(\phi)$, which agrees with what we found before.

To compute the scale factor along a meridian for a cylindrical projection of S_f, we compare the length of a small arc of a meridian with the length of its image on the map. On the map, we are just looking at the difference in the heights of two nearby horizontal lines. This distance is therefore dh. On the surface itself, the length of a small arc of a meridian is given by the arclength element $ds = \sqrt{1 + (f'(x))^2}\, dx$. Thus, the scale factor along the meridian, denoted by M_m, is equal to

$$(55) \qquad M_m = dh/ds = \frac{h'(x)}{\sqrt{1 + (f'(x))^2}}.$$

For an azimuthal projection, the difference in the radii of the circles representing two nearby parallels is dr, so we get

$$(56) \qquad M_m = \frac{r'(x)}{\sqrt{1 + (f'(x))^2}}$$

for these maps. (If the function r is decreasing, then we take the absolute value of this.)

On the sphere of radius 1, where $f(x) = \sqrt{1-x^2}$ and $x = \sin(\phi)$, we get that

$$\frac{1}{\sqrt{1 + (f'(x))^2}} = \sqrt{1 - x^2} = \cos(\phi) = \frac{dx}{d\phi}.$$

Hence, either $M_m = h'(x)(dx/d\phi) = h'(\phi)$, for a cylindrical projection, or $M_m = r'(x)(dx/d\phi) = r'(\phi)$ for an azimuthal projection. This agrees with what we found in Chapter 3.

4. Conformal mappings

A mapping of the surface S_f is *conformal* if the images of any two intersecting paths on the surface intersect at an angle equal to that between the original paths themselves. Just as with the sphere (see Chapter 9), a conformal mapping of S_f must show the parallels at right angles to the meridians and satisfy the condition $M_p = M_m$ at every point. Moreover, a cylindrical or azimuthal projection of S_f that satisfies $M_p = M_m$ at every point will be conformal.

For a cylindrical projection of a surface of revolution S_f, as defined above, the condition for conformality, that $M_p = M_m$, becomes

$$\frac{1}{f(x)} = \frac{h'(x)}{\sqrt{1 + (f'(x))^2}},$$

or, in other words,

$$h'(x) = \frac{\sqrt{1 + (f'(x))^2}}{f(x)}$$

on the interval I. If we choose a suitable reference point c in the interval, then integrating the previous equation gives us the formula

$$(57) \qquad h(x) = h(c) + \int_c^x \frac{\sqrt{1 + (f'(t))^2}}{f(t)} \, dt$$

for x in the interior of I.

For the azimuthal projections of S_f, the condition that $M_p = M_m$ says that

$$\frac{r(x)}{f(x)} = \frac{|r'(x)|}{\sqrt{1 + (f'(x))^2}}.$$

We can rewrite this as

$$\frac{|r'(x)|}{r(x)} = \frac{\sqrt{1 + (f'(x))^2}}{f(x)}.$$

Integrate both sides and solve for $r(x)$, again selecting a suitable reference point c, to get

$$(58) \qquad r(x) = r(c) \exp\{\pm \int_c^x \frac{\sqrt{1 + (f'(t))^2}}{f(t)} \, dt\}$$

for x in the interior of I, where the $\pm$ in the formula depends on whether $r(x)$ is increasing or decreasing.

As an exercise, verify that, for the sphere of radius 1, where the profile curve is $f(x) = \sqrt{1 - x^2}$ for $-1 \leq x \leq 1$, the function $h(x)$ defined in (51) for the Mercator projection satisfies the equation (57), with $c = 0$ and $h(0) = 0$. Also, verify that the function $r(x)$ defined in (52) for the stereographic projection satisfies the equation (58), with $c = 0$ and $r(0) = 2$. Any other value for $r(0)$ yields a rescaled version of the stereographic projection.

A close look at the formulas (57) and (58) shows that, if $r(x)$ is assumed to be increasing and if $r(c) = e^{h(c)}$, then $r(x) = \exp(h(x))$. See [Feeman, 2001] for some interesting implications of this relationship between h and r.

4.1. A technicality. As a somewhat technical matter, observe that if $f(a) = 0$, where a is an endpoint of the interval I, then the meridians of S_f meet at $(a, 0, 0)$, which we can think of as a pole on S_f, much like the north and south poles on the earth. The integral in formula (57), with $x = a$, is, in this case, improper and diverges to $\pm\infty$. For the conformal cylindrical projection, this means that the pole cannot be shown on the map. This makes sense when we realize that the meridians on the cylindrical map are vertical lines and, hence, don't meet. For the conformal azimuthal projection, it would seem that the pole at $(a, 0, 0)$ could be shown on the map as the point at the origin, provided that the integral in (58) diverges to $-\infty$. However, if the limit $\lim_{x \to a} f'(x) = m$ is finite in value, then S_f can be approximated near the pole by a cone of slope m. A cone can be conformally slit open and laid out flat to form not a full disc but a sector of a disc. The total angle formed by the meridians of a cone is less than a full 2π radians. Thus, a conformal azimuthal projection of S_f can include the pole only if $\lim_{x \to a} f'(x) = \infty$. In this case, the surface S_f will have a well-defined tangent plane at its pole onto which it can be "projected" to form the azimuthal map.

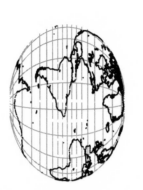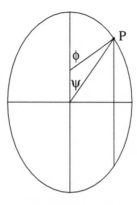

FIGURE 1. On the left is a portrait of the earth as an oblate spheroid; the profile curve is an ellipse. On the right we see that the angle ψ of a parallel away from the equator is different than the angle ϕ made by the sun at noon on the equinox.

5. Examples

5.1. Oblate spheroid. As a first application, let's look not at an "other" world, but at another model of our own world – the oblate spheroid or ellipsoid. The profile curve for this surface is half of an ellipse, defined by the function $f(x) = (b/a)\sqrt{a^2 - x^2}$, where a and b are positive numbers. (For the earth, a corresponds to the polar radius and b to the equatorial radius. The poles are at the points $(\pm a, 0, 0)$ and the equator is the parallel at $x = 0$.) In this case, the arclength element is given by

$$\sqrt{1 + (f'(x))^2}\, dx = \left(\frac{a^4 + (b^2 - a^2)x^2}{a^2(a^2 - x^2)} \right)^{1/2}.$$

Integrating the formula (57) and setting $h(0) = 0$, we find that, for a cylindrical projection of the ellipsoid onto a rectangle to be conformal, the height function $h(x)$ for the parallels must satisfy

$$(59) \qquad h(x) = \int_0^x \frac{\sqrt{1 + (f'(t))^2}}{f(t)}\, dt = \int_0^x \frac{\sqrt{a^4 + (b^2 - a^2)t^2}}{b(a^2 - t^2)}\, dt.$$

This is actually a rather horrid expression in the end, involving both natural logarithms and the arctangent function. It does, however, reduce to what we found for the Mercator projection if we take $a = b = 1$.

For a conformal azimuthal projection of the spheroid, the formula (58) yields a radius function $r(x)$ that is likewise intractable. Despite being complicated, the expressions for $h(x)$ and $r(x)$ can still be evaluated numerically for selected values of x. Therefore, we can still succeed in making maps of the oblate spheroid.

One of the difficulties in working with an ellipsoid instead of a sphere involves latitude. On the sphere, the angle that a parallel makes with the equator coincides with the angle that the sun makes with a vertical gnomon at noon on the equinox. On an oblate spheroid, these two angles are different, as illustrated in Figure 1. (See [Cotter, 1966].) As we mentioned in Chapter 1, it is easier in practice to use the wealth of map projections available for the sphere. Then, if the utmost accuracy is needed, a flat map of the sphere can be composed with a map of the spheroid onto

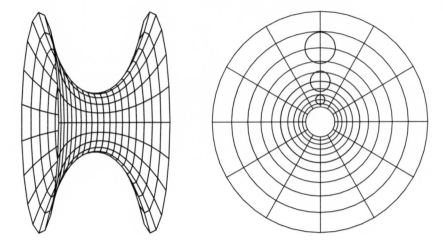

FIGURE 2. On the left is a catenoid; the profile curve is a catenary. On the right is the base grid for a conformal azimuthal map of part of a catenoid; the small circles show the pattern of area distortions of the map.

the sphere. Not surprisingly, a variety of maps from the ellipsoid onto the sphere have been developed, including conformal and equal-area maps. See [Bugayevskiy and Snyder, 1995] for more details.

5.2. Catenoid. A chain hanging under the influence of gravity describes a curve called a *catenary* (from the Latin word for "chain"). It is a result of several Bernoullis that, when extended to an infinite curve, the catenary is the graph of the function $f(x) = \alpha \cosh(x/\alpha)$, where $\alpha = f(0)$ is a positive constant. Revolving this curve about the x-axis produces a surface of revolution called a *catenoid*. (See Figure 2.) The catenoid is a minimal surface, formed, for instance, by a soap film extended between two circular wire frames in parallel planes.

Intersecting the catenoid with any plane perpendicular to the x-axis produces a parallel while the meridians are obtained by intersecting the catenoid with any plane (in xyz-space) that contains the x-axis. Each meridian looks like a copy of the catenary $y = f(x)$. Each parallel is a circle, of radius $f(x)$, that is perpendicular to every meridian. The arclength element along the catenary is $ds = \sqrt{1 + [f'(x)]^2}\, dx = \cosh(x/\alpha)\, dx$.

To construct a conformal azimuthal projection of a catenoid, we first compute that $\sqrt{1 + (f'(x))^2}/f(x) = 1/\alpha$. According to the formula (58), the appropriate radius function for the map is then given by

$$(60) \qquad r(x) = r(0) \exp\{ \int_0^x \frac{1}{\alpha}\, dt \} = r(0)\, e^{x/\alpha}.$$

If we take $r(0) = \alpha$, then the parallel at 0 will be a standard line for the map; that is, the map will show this parallel at its true length, $2\pi\alpha$. In this case, we get $r(x) = \alpha e^{x/\alpha}$ for all real numbers x. Figure 2 shows the resulting grid for the portion of the catenoid between $-1 \le x \le 1$, with $\alpha = 1$.

FIGURE 3. A conformal map of the earth's surface onto a catenoid.

A conformal cylindrical projection of the catenoid is obtained by applying the formula (57). With $h(0) = 0$, we get, for arbitrary x,

$$h(x) = \int_0^x \frac{1}{\alpha}\, dt = \frac{x}{\alpha}.$$

We can make an interesting conformal map of the earth's surface onto a catenoid by first applying the Mercator projection to get a flat map of the earth, then using the inverse of the conformal cylindrical projection of the catenoid to get from the flat paper back to the catenoid. Figure 3 shows the result (or actually an orthographic projection of it onto the paper).

5.3. Pseudosphere. The pseudosphere, also known as the bugle surface or tractroid, is generated by a curve, called the *tractrix*, that starts at the point $(0, 1)$ and moves so that its tangent line always reaches the x-axis after running for a distance of exactly 1. (See Figure 4.) Analytically, this condition is described by the differential equation $f'(x) = -f(x)/\sqrt{1 - [f(x)]^2}$, for $x > 0$, with $f(0) = 1$. The

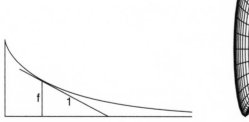
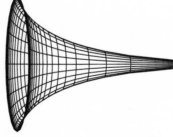

FIGURE 4. The tractrix, shown on the left, is determined by the condition $f' = -f/\sqrt{1 - f^2}$. The pseudosphere, on the right, is generated by the tractrix and is a surface of constant negative Gaussian curvature.

arclength element along this curve is $ds = \sqrt{1 + [f'(x)]^2}\, dx = -[f'(x)/f(x)]\, dx$. Upon revolving the tractrix about the x-axis, every value of x corresponds to a parallel of the bugle surface while each meridian is a copy of the tractrix.

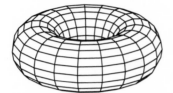 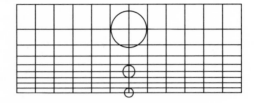

FIGURE 5. The torus with $R_1 = 1$ and $R_2 = 2$ is shown on the left. On the right is the base grid for a conformal cylindrical map of the torus with $R_1 = R_2 = 1$. The grid shows latitudes $0 \leq \phi \leq 3\pi/4$. The small circles show the area distortion pattern of the map.

Therefore, a cylindrical projection whose height function, $h(x)$, is increasing and satisfies $h(0) = 0$ has scale factors $M_p = 1/f(x)$ and $M_m = dh/ds = -h'(x)f(x)/f'(x)$. Conformality is obtained by taking

$$h(x) = \int_0^x \frac{\sqrt{1 + [f'(t)]^2}}{f(t)}\, dt = \int_0^x \frac{-f'(t)}{(f(t))^2}\, dt = \frac{1}{f(x)} - 1.$$

For a conformal azimuthal projection, let us take the radius function $r(x)$ to be decreasing and satisfy $r(0) = 1$. The scale factors are $M_p = r(x)/f(x)$ and $M_m = dr/ds = -r'(x)f(x)/f'(x)$. Setting $M_p = M_m$ yields

$$r(x) = \exp\left\{ 1 - \frac{1}{f(x)} \right\} = \exp\{-h(x)\},$$

for $x > 0$. This maps the pseudosphere conformally onto the punctured unit disc in the plane.

5.4. Torus. A slightly different approach is needed to handle the torus, which can't be obtained by revolving the graph of a function. Instead, choose two positive numbers, R_1 and R_2, satisfying $0 < R_1 \leq R_2$. The circle *lying in the yz-plane* with radius R_1 and center at $(0, R_2, 0)$ can be parametrized by $(0, R_2 + R_1 \cos(\phi), R_1 \sin(\phi))$, for $-\pi \leq \phi \leq \pi$. Revolving this circle about the z-axis generates a torus in xyz-space parametrized by the equations

(61)
$$\begin{aligned} x &= (R_2 + R_1 \cos\phi) \cdot \cos\theta, \\ y &= (R_2 + R_1 \cos\phi) \cdot \sin\theta, \\ z &= R_1 \sin\phi, \end{aligned}$$

for $-\pi \leq \phi, \theta \leq \pi$. Each fixed value of ϕ corresponds to a parallel, a circle of radius $[R_2 + R_1 \cos(\phi)]$. Since fixing a value of ϕ also fixes the value of z, the parallels are the horizontal cross-sections of the torus. Each meridian is a circle of radius R_1 obtained by intersecting the torus with a plane containing the z-axis. This is the same as fixing the value of the angle θ in the parametrization. We can think of the parallel for $\phi = \pm\pi$, with radius $R_2 - R_1$, as the "inner equator" of the torus and the parallel at $\phi = 0$, with radius $R_1 + R_2$, as the "outer equator".

For a cylindrical projection of the torus, we map the parallel at ϕ to a horizontal line segment of length 2π at height $y = h(\phi)$, where h is an increasing function satisfying $h(0) = 0$. The images of the meridian at angle θ is all or part of the vertical line $x = \theta$. The scale factors are then $M_p = 1/[R_2 + R_1 \cos(\phi)]$ and, since

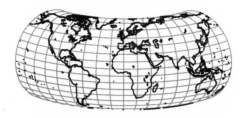

FIGURE 6. A conformal map of the earth onto part of a torus.

the arclength element on each meridian is $ds = R_1 d\phi$, $M_m = dh/ds = dh/(R_1 d\phi)$. For a conformal map, we require that $M_p = M_m$. That is,

$$h(\phi) = \int_0^\phi \frac{R_1}{R_2 + R_1 \cos(t)} dt.$$

In the case where $R_1 = R_2$, so that the "inner equator" of the torus is just the point at the origin, then we get simply $h(\phi) = \tan(\phi/2)$. Figure 5 shows the base grid for latitudes $\phi = 0$ at the bottom through $\phi = 3\pi/4$ at the top.

An azimuthal projection of the torus will show the parallel at ϕ as a circle, centered at the origin, of radius $r = r(\phi)$ and the meridian at θ as all or part of the ray at angle θ. Assuming that the function $r(\phi)$ is increasing, the scale factors are $M_p = r(\phi)/[R_2 + R_1 \cos(\phi)]$ and $M_m = dr/(R_1 d\phi)$. Therefore, we will have a conformal azimuthal projection of the torus provided that

$$r(\phi) = r(0) \cdot \exp\left\{ \int_0^\phi \frac{R_1}{R_2 + R_1 \cos(t)} dt \right\}.$$

The integrand is the same as that encountered in the conformal cylindrical projection.

Figure 6 shows an orthographic view of a conformal map of most of the earth's surface onto part of a torus with $R_1 = 1$ and $R_2 = 2$. The map was created by first forming a scale-model of the Mercator projection of the earth and then applying the inverse of the conformal cylindrical projection for the torus.

6. Equal-area maps

To produce an equal-area cylindrical or azimuthal map of a surface of revolution, we need to have $M_p \cdot M_m = 1$. You are invited to use this to construct area preserving cylindrical and azimuthal maps for the examples above or for your own favorite surfaces of revolution.

7. Distortion analysis

As with the sphere, we can use the values of $M_p \cdot M_m$ and M_p/M_m to measure the distortions in area and angle for cylindrical and azimuthal projections of a surface of revolution. Figures 2 and 5 show examples of Tissot's indicatrix for two conformal maps.

APPENDIX A

Aspects of thematic cartography: Symbolization, data classification, and thematic maps

1. Map classification

One of the first things that we might do in order to sort out the vast quantity of maps that exist is try to determine some ways that we can loosely *classify* them or distinguish among them.

1.1. General purpose *vs.* thematic maps. Because every map is produced by its author with some purpose in mind, a first step in classifying maps is to distinguish them according to the *principal purpose* for which the map was compiled.

Maps that have as their principal goal to show locations and names of specific places and things are called **general purpose maps**. For example, a typical road maps is a general purpose map — it shows the locations of roads, lakes, rivers, airports, towns, and so on. Some general maps include certain information that requires data collection, such as elevations of mountains or populations of cities. But communicating this information is *not* the primary purpose of the map.

On the other hand, **thematic maps** are those intended *primarily* to tell us something about the distribution of some kind of stuff. Thematic maps are those that *require* the collection, analysis, and classification of data. For example, a thematic map might show population density, the distribution of farmland, the comparative gross national products of various countries, the circulation of newspapers, the distribution of ethnic groups, or the rates of infant mortality or adult literacy.

The classification of a map as either general or thematic is based on the primary purpose for which the map was compiled. Often a map will have elements of both types, in which case a decision must be made about the map's intended aim. For instance, topographic maps are usually classified as general purpose maps, although they do include a number of thematic elements, such as elevation, forest cover, and urban areas.

1.2. Large *vs.* small scale. We can also distinguish various maps based on their **relative scale**.

A map that has a large scale factor or representative fraction can only represent a small portion of the earth's surface. These are classified as **large scale maps**. Conversely, a map having a small scale factor or representative fraction is a map of large area. These are classified as **small scale maps**.

As a rough rule, a representative fraction of 1:30 000 000 or smaller is considered to be a small scale. Representative fractions of 1:24 000 or larger are considered large scales. Scales between these two are usually referred to as *intermediate*. For example, a map of a university campus on a letter-sized sheet of paper is a large

97

scale map. It depicts a relatively small area of the earth's surface. On the same letter-sized piece of paper, a map of the entire world is a small scale map. It represents a large portion of the the earth's surface (all of it!).

On a large scale map, things appear large, because it is a map of a small area. On a small scale map, things appear small, because the area being mapped is very large. For example, on a large scale map of a town, a road might be illustrated as a thick line or as two parallel lines. In contrast, a small scale map of the earth's eastern hemisphere would not show roads at all due to their inconsequential size relative to the sizes of the land masses being depicted.

By classifying maps into the categories of general *vs.* **thematic or of large scale** *vs.* **small scale, we actually obtain a total of four categories - small scale general, large scale general, small scale thematic, and large scale thematic.**

1.3. Refining the classification process. The classification of a map as general or thematic is based on the map's primary purpose. We can refine this classification by grouping maps together based on their specific purpose or use.

For instance, we might use a classification system that separates out road maps from navigational and aeronautical charts (all of which are maps used for the purpose of way-finding). Depending upon how specific we want our classification system to be, we can continue to classify maps for more and more narrowly defined purposes.

Thus, there are three main ways in which all maps can be classified - as being general purpose or thematic, as large or small scale, and according to the specific purpose for which the map was made.

The rest of this chapter provides an overview of some of the principal aspects of thematic maps.

2. Map symbolization

Using a variety of colors to distinguish neighboring countries, using a blue line to indicate the path of a river, and using a picture of an airplane to mark the location of an airport are all examples of **map symbolization**. That is, they are all examples of how symbols are used on maps to highlight certain of the map's features. The choices of what features to symbolize and what symbols to use are closely linked to the map's *purpose.*

Before we see how map symbolization is related to displaying data on a map, let us look at the three basic types of symbols.

Map symbols can be classified as **point**, **line**, or **area** symbols according to the nature of the feature being highlighted.

2.1. Point symbols. Point symbols are used to distinguish map features located at specific points on the map. Such features might include cities on a small scale map of a continent or large country; airports, cemeteries, and schools on a large scale map of an urban area; important industrial sites on a map showing patterns of trade or commerce; the location of the buried gold on a pirate's treasure map.

2.2. Line symbols. Map features whose location follows a path along the earth's surface can be highlighted using **line symbols**. Railroad lines, highways for automobiles, rivers and streams, trade routes, wind and ocean currents, and political boundaries are a few examples. The symbols used to represent such features generally follow (approximately) the path occupied by the feature itself. Thus, a line with cross hatches might indicate a railroad line, a red line might indicate a highway, a blue line a river, a pink arrow the direction of trade in iron from an exporter to an importer, a curved arrow an ocean current and its direction.

2.3. Area symbols. A map feature that occupies a significant amount of area on a given map would be represented by an **area symbol**. For instance, a map might show the dominant vegetation of different regions of Africa by shading desert regions in brown, grasslands in yellow, and forested areas in green. Colors might also be used to distinguish different countries on a political map or to illustrate elevation above sea level on a topographic map. A large scale map of Portugal might use an area symbol to indicate major urban centers. Area symbols usually involve some sort of shading pattern.

3. Data classification

Open any modern atlas and you will see a wealth of maps displaying data of all kinds, such as world distribution of temperature, seasonal precipitation, vegetation, population density, world languages, diet, life expectancy, types of agriculture, production and distribution of copper, energy consumption, and the world's refugee population. Maps can be designed to illustrate the spreads of diseases, the migration of butterflies, the distribution of fire stations, voting patterns in an election, and the amount of money spent on education. In short, there is essentially no limit to the kinds of data and information maps can communicate.

Before we can design a map that displays data, we must first determine what general type of data we have and what sort of map symbols can best be used to display them.

3.1. Nominal, ordinal, and interval data. Suppose we were to describe the group of people who lived in our town or neighborhood.

We could categorize each person according to his or her occupation – teacher, student, builder, hospital worker, and so on. We could categorize our neighbors by gender. These are examples of **nominal data** categories. Each category (occupation, gender) is divided into data classes (teacher/student/etc., female/male) each of which has a name. There is no attempt to impose any kind of hierarchy or order on these classes by saying that one class is somehow bigger or better than another.

Suppose instead we categorize the students in the class by hair color, using three data classes – light hair, medium hair, dark hair, or by height – short, average, tall. Now there is an implied order or hierarchy among the data classes, but there is no attempt to provide specific numerical divisions between the classes. These are examples of **ordinal data** categories.

By using specific numerical divisions between data classes of an ordinal data category, we can create an **interval data** category. For instance, if we take the category of height and classify each person as 'not taller than five feet', 'taller than five feet but not taller than six feet', or 'taller than six feet', then we have given specific numerical definitions to the terms 'short', 'average', and 'tall'. Each data class is now a specific interval of heights.

When using interval data, *the different class intervals must not overlap.* Overlapping data class intervals create confusion about how to treat data that lie in more than one interval.

Nominal, ordinal, and interval data categories are the three main types used on maps. A fourth type, **ratio data**, is a variation on an interval data category where each interval is an interval of ratios or percentile ranks.

3.2. How data and map symbols work together. We have already discussed the three main types of map symbols – point, line, and area symbols, and the three main categories of data – nominal, ordinal, and interval. In order to actually display a given data set on a map, we must match an appropriate set of map symbols to the given type of data. Here are some examples.

3.2.1. *Using point symbols with data.*

(1) Suppose that we label the towns and cities on a map using the following symbols: a star ($\star$) to designate a state capital, a triangle (Δ) to designate a county seat, and a dot ($\bullet$) for all other towns. The data category is nominal and each data class is assigned a point symbol. There is no attempt to impose a hierarchy on the data or on the symbols. This is **nominal point data**.

(2) Suppose we again label the towns on a map, this time designating each town as either a 'village', a 'town', or a 'city', with no specific numerical definitions of these terms. To symbolize these classes, we could use a small dot for a village, a larger dot for a town, and a still larger dot for a city. Now there is a hierarchy to the data classes and this hierarchy is mirrored in the symbols. This is **ordinal point data**.

(3) Thirdly, suppose we define three classes of towns according to specific population intervals, such as 'not more than 10 000 residents', 'more than 10 000 but not more than 40 000 residents', and 'more than 40 000 residents'. Assign a small dot to designate a town in the least populous interval, a larger dot for the next most populous interval, and a still larger dot for the most populous interval. In this case, there is a hierarchy on the data as well as on the symbols and the data classes are defined by specific intervals. This is **interval point data**.

3.2.2. *Using line symbols with data.*

(1) A map that shows state borders using red lines, rivers using blue lines, and roads using black lines is displaying **nominal line data**.

(2) A map that shows roads using the classes 'unpaved road', 'paved road', and 'divided highway', with a dashed line used to symbolize an unpaved road, a solid black line to designate a paved road, and two solid black lines to designate a divided highway is displaying **ordinal line data**.

(3) A map might use arrows of different widths to indicate the flow of steel from one country to another. If the width of the arrow indicates the amount of steel involved according to certain specific intervals, then the map is displaying **interval line data**. The contour lines on a topographic map, each indicating a certain specific elevation above sea level, provide another example of interval line data.

3.2.3. *Using area symbols with data.*

(1) A topographic map might show regions of swamp, grassland, and forest using clumps of marsh reeds for the swamps, tufts of grass-like blades for the grasslands, and clusters of trees for the forests. This is an example of **nominal area data**. There are three data classes (swamp, grassland, forest) with no order or ranking imposed on them.

(2) A map showing areas of relatively low elevation above sea level in green, areas of intermediate elevation in yellow, and areas of relatively high elevation in brown, without assigning numerical values to the terms 'low', 'intermediate', and 'high', would be displaying **ordinal area data**.

(3) Finally, suppose we separate the category of elevation above sea level into three data class intervals – elevations of not more than 150 meters above sea level, elevations of more than 150 meters but not more than 300 meters, and elevations of more than 300 meters above sea level. We might then shade various regions on our map in green, yellow, or brown, respectively, according to the average elevation over the region. This would be an example of **interval area data**.

4. Types of thematic maps

Having discussed the different types of map symbols and data categories, we will now look at some of the common approaches cartographers take to actually displaying the data.

4.1. Dot maps. Perhaps the simplest way to indicate the location on a map of a particular object is to mark the location with a dot. For instance, on a map of the United States, we might mark every state capital with a dot. It would not be practical, however, to mark every farm with a dot, or every sheep with a dot, or every acre of harvested cropland with a dot. So, instead, we might *assign a specific numerical value to each dot.*

Let's suppose that we decide to use dots to show the distribution of sheep around the world. We might decide that one dot will represent 200 000 head of sheep. Placing the appropriate numbers of dots in each geographical region of the world results in the map on the top of page 44 of Goode's World Atlas [Espenshade, et al., 1995]. This is a standard example of a **dot map**. In principal, one ought to be able to determine the exact number of sheep in each region by counting the dots, but one look at the map tells you that that is not practical. Rather, as Michael R. C. Coulson has put it ([Coulson, 1990]),

> the power of the dot map is in the overall pattern of the distribution
> that is revealed.

To prepare a dot map, the cartographer must give some consideration to the way in which people perceive dots. The size of each dot as well as its assigned numerical value should be chosen so that the overall balance of dot size to dot density is within the range that people can successfully interpret visually. The *nomograph* is a special tool, developed on the basis of research into dot perception, that helps the cartographer select appropriate dot sizes and values.

4.2. Proportional point symbols. Two symbols commonly used in thematic maps to display interval (or ratio) point data are *nested graduated circles* and *pie charts*.

For instance, the map on the top of page 79 in Goode's World Atlas [Espenshade, et al., 1995] shows both the size and composition of the labor force in various cities in the United States. The first data category, size of the labor force, is divided into data class intervals. Each interval is assigned a different size of circle to represent it. In the map's legend, these circles are shown nested inside each other. Hence, the term *nested graduated circles*. Each city included in the study will be marked with a circle of the appropriate size.

The second data category, composition of the labor force, is divided into what are actually nominal data classes – services, manufacturing, transportation and communication, and so on. Each of these is assigned a color. The circle located at a given city is then filled in with a pie chart showing the proportion of the labor force in that city that belongs to each class.

In selecting the sizes of the graduated circles, the cartographer must take into account the fact that it is the *area* of the circle, not its diameter, that will impress the map user. Because the area of a circle grows like the square of the diameter, it follows that the ratio of the diameters of circles representing two different data class intervals should be like the square root of the ratio of the numerical values of the intervals.

4.3. Flow maps. A straightforward way to indicate linear movement between places on a map is to place an arrow or a curve along the path of the movement from one place to the other. For instance, on weather maps, the movement of the jet stream is indicated by an arrow showing the direction and path of the current. Airlines often publish maps that show the major destinations the airline services, using curves to indicate the flight paths between destinations. (Since one could presumably fly in either direction, it would not make sense to use an arrow.) In general, maps that show this sort of movement are called **flow maps**.

Flow maps are widely used for statistical purposes to show not only some type of movement between places but to show the *amount* of movement as well. This is achieved by using arrows of various widths, where the width of the arrow is keyed to a numerical value or an interval of numerical values that gives the amount of movement. Such maps are particularly useful in economic geography.

For example, let's suppose that we decide to use a flow map to show selected major movements of steel between certain countries. We will use arrows, also called *flow lines*, whose width is proportional to the tonnage of steel flowing along a given route. Given the amounts of steel involved, we might decide to take one-half millimeter of arrow width to represent one million metric tons of steel. The result would be much like the map on the top of page 49 of Goode's World Atlas [Espenshade, et al., 1995].

One obvious issue in the design of a flow map is the choice of numerical values for the various arrow widths. A frequently occurring value should not be assigned too small of a width to really be noticed. On the other hand, a large range of numerical values could result in arrows so wide that they block out important parts of the bakground map. There are other important design issues as well that we will not go into here.

4.4. Choropleth maps. We turn now to three types of thematic maps that involve shaded areas – choropleth and isometric maps and cartograms.

A choropleth map is used to display data that is collected and aggregated within various regions, such as counties or states within the United States. Each region is shaded according to where the aggregate data for that region fits in the data classification scheme.

For example, it is common after a presidential election in the United States to see maps in which each state is shaded according to which candidate received the majority of the vote in that state. For these maps there are only as many data classes as there are major candidates. A potentially misleading aspect of such maps is that one candidiate might win many states but only by a narrow margin. The actual voting percentages are not shown, so a close race might look like a landslide election. Also, some states with large land area have relatively small voting populations. Winning such a state looks more significant on the map than it really is in terms of actual votes.

A more refined, and more typical, example of a choropleth map is found on the bottom of page 79 of Goode's World Atlas [Espenshade, et al., 1995]. This map shows the value added by manufacture for each county in the United States. The range of values for the data has been separated into data class intervals each of which has been assigned a color. Each county is shaded according to the interval in which its aggregate data falls. Notice that the colors are graduated in shades of yellow to orange to red. In this way the map reader can visualize the progression in the data – a darker shade corresponds to more value added.

More choropleth maps are shown on pages 75–76 of Goode's World Atlas [Espenshade, et al., 1995].

In a choropleth map, each region is shaded according to the aggregate data for the region. Data for individual places within a region are not shown. In this way, a choropleth map differs from an isometric map.

4.5. Isometric maps. Think of a weather map of a large region that shows the current temperature at each location. In such a map, the range of temperatures is usually separated into intervals and each point is colored according to the interval in which its current temperature lies. The range of colors might be from a dark blue for colder temperatures to yellow then orange and red for hotter temperatures. Also, it is implied that the temperature at a point lying on the boundary of, say, a red region and an orange region has as its temperature the value that defines the upper temperature for the orange interval and the lower temperature for the red interval. In other words, each line that separates one color region from another on the map has a specific value. These boundary lines are called **isolines** for the map. This map is an example of an **isometric map**. See, for instance, the world temperature maps on pages 12–13 of Goode's World Atlas [Espenshade, et al., 1995]. A map depicting elevation above sea level is another example of an isometric map. The isolines, called contour lines in this case, show elevation in regular increments.

An isometric map differs from a choropleth in two essential ways. First, unlike a choropleth map, the data for the isometric map are not aggregated over certain regions. Instead, each point is colored according to its own value. Second, choropleth maps have no isolines since the data are aggregated and boundaries between regions have no specific implied value.

To create an isometric map, it is essential that the data vary continuously over the region to be mapped. At least it must be reasonable to think of the data as varying continuously. Discrete data won't work since then the boundary between two regions won't have a specific value. To make an accurate isometric map, the cartographer must have a detailed knowledge of the data distribution since every point must get the appropriate shading.

4.6. Cartograms. Consider a map of the United States that shows a state colored in blue if the majority of that state's vote in the last presidential election went to the Democratic candidate and colored in red if the majority of the vote went to the Republican candidate. (This is an example of a choropleth map.) One misleading feature of this map (there are several) is that it does not show the actual number of votes each candidate received. Instead, the visual impression of the colors dominates. Thus, a majority vote for one candidate in a state that has a large land area but a small population (e.g., Wyoming) makes a big visual impact while a majority for the other candidate in a small state with a large population (e.g., Massachusetts) makes little visual impression. So the map doesn't necessarily tell us what is going on.

One way to address this problem is to alter the depiction of the states on the map so that each state is shown at a size that is proportional to its population, rather than to its land area. A state with twice as many people as another state will be shown twice as large under this scheme. In this way, the visual impact of each state's coloration is more in keeping with the impact of its numerical vote in the election.

This technique of rescaling the sizes of different regions on the map relative to a measured value, like population, results in a type of map known as a **cartogram** or a *value-by-area map*.

As another example of a cartogram, the map at the top of page 31 in Goode's World Atlas [Espenshade, et al., 1995] shows the population per physician around the world. If each country were shown at its correct proportional size in terms of land area, then a country with a large area but a small population would have a visual impact out of proportion with what the data represent. In the case of Canada, for instance, the data value for population per physician is low. Showing Canada at its true size would make the overall availability of physicians worldwide seem less of a problem than it really is. The solution is to rescale the size of each country on the map in relation to its population, rather than its land area. Then the map is visually dominated by the availability of physicians in the countries with the largest populations. In this way the map is better able to communicate its intended message.

Certain design problems arise with cartograms, with the most obvious being that the borders between countries or regions often have to be altered in order to resize everything. Gaps and other sorts of mismatches between neighboring regions invariably occur. Overall shapes of countries and continents get badly distorted, so the cartographer must struggle to keep things looking somewhat "right" while still resizing regions appropriately. This difficulty has given rise to two types of cartograms – *contiguous* and *non-contiguous*.

In a contiguous cartogram, shapes of regions are distorted, often severely, but the connection between the map and true geographical space is maintained. However, the distortions can be confusing and disorienting for the map reader. More examples of contiguous cartograms are shown on pages 30–31 of Goode's World Atlas [Espenshade, et al., 1995].

In a non-contiguous cartogram, each region is shown separately, floating on its own as it were. While each region is resized appropriately, its boundary no longer has to match up with those of its neighboring regions. Thus, each region can be shown with its correct shape. On the other hand, the discontinuity of the map obscures the connection to geographical reality. Also, the true geographical shape of the overall region shown in the map often gets badly distorted.

APPENDIX B

Laboratory Projects

1. Project: Historical geodesy

In this project, you will determine, mainly by direct measurement, the latitude and longitude of a particular place on earth. **You will need the following tools:** (i) a straight object of known length, such as a meter stick, to use as a gnomon; (ii) a clock or watch accurately set to Greenwich Mean Time (GMT); (iii) a tape measure or other device for measuring lengths of things; (iv) a protractor; (v) a piece of string; (vi) some chalk or other marker; (vii) a calculator; (viii) an analemma (see section 1.1 below); (ix) a friend or two. **First read through the entire procedure for the project.** Then carry out the following steps in the order given. Be careful and patient – it will improve your accuracy, understanding, and enjoyment. **Have fun! Let your friends help and learn with you!**

(1) Set a reliable clock or watch to Greenwich Mean Time (GMT).

(2) Sometime before your location's local solar noon, on a sunny day, go outside and place a gnomon of known height vertically into the ground. An existing pole or tree may be used if you can reach the top of it and know its height.

(3) Starting before your location's local solar noon, begin tracking the shadow of the gnomon. At frequent intervals, mark the tip of the shadow and record the GMT time. Continue until you are sure that the shadow is getting longer again.

(4) Calculate the time of your location's *local solar noon* by determining the time on the GMT watch at which the shadow cast by your gnomon was the shortest. Also, measure and record the *length* of this shortest shadow. (For instance, mark the tip of the shadow with chalk and measure the distance of the mark from the base of the gnomon.)

(5) Determine the angle between the gnomon and the sun's rays at local solar noon, as shown in Figure 1, by holding a string between the top of the gnomon and the tip of the gnomon's shortest shadow and using a protractor.

(6) As a check on the angle measurement you made in the last step, you can also compute the angle between the gnomon and the sun's rays at local solar noon as follows. As shown in Figure 1, let h be the height of the gnomon and s the length of its shadow at local solar noon. Then the tangent of the angle t is given by $\tan(t) = s/h$. Compute the ratio s/h and then determine the value of $t = \arctan(s/h)$ using a calculator.

(7) Now use the analemma to determine the latitude ϕ at which the sun is directly overhead for the day on which you are taking your measurements. The angle t that you computed in the previous steps is the difference in

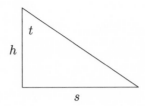

FIGURE 1. angle between the gnomon and the sun's rays

latitude between your location and the latitude ϕ where the sun is directly overhead. **Determine the latitude of your location.** Don't forget to take north/south into account.

(8) The analemma also records the difference between *local solar time* and *local mean time* for each day of the year. Use the analemma to determine this difference for the day of your measurements. (Where the analemma chart indicates that the sun is "slow", that means that local solar noon occurs after local mean noon; when the sun is "fast", that means that local solar noon occurs before local mean noon.) Now use this difference, together with the time at which your gnomon's shadow was shortest (this is the local solar noon you computed in step 4), to compute the correct Greenwich Mean Time of your location's *local mean noon*.

(9) Now **determine the longitude of your location** by computing the difference between 12:00 GMT and the Greenwich Mean Time of your location's local mean noon (which you just computed in step 8). Convert this time difference to a longitude angle by the relations shown in the table. Don't forget to include East/West.

Time unit	longitude angle
1 hour of time	15° of longitude
1 minute of time	15′ of longitude

1.1. The analemma. The *analemma* is a graphic tool, shaped like a figure-eight and pictured on many globes, that is used to help in the determination of latitude. It indicates the parallel of latitude, between the Tropics of Cancer and Capricorn, at which the sun shines directly overhead at local solar noon on any given day of the year. To use it, select a date on the figure-eight shaped curve of the analemma. The parallel on which that date is placed corresponds to the latitude at which the sun is directly overhead on that date.

A second function of the analemma involves the measurement of time. Even though we say that every day has twenty-four hours, the length of the solar day is not constant throughout the year. This is due partly to the fact that the earth does not move around the sun at a constant speed and partly to certain effects of the tilt of the earth's axis. Our clocks are actually set according to an averaging out of solar time, called *mean time*. Recorded across the top of the analemma is the difference between our local solar time and our local mean time for each day of the year. For instance, on the Fourth of July, the analemma indicates that solar time is four minutes slower than mean time. With our clock set to the mean time for our time zone, suppose we find that the sun is at its highest point in the sky at 12:03 pm. Since the sun is slow by four minutes on that day, we conclude that our mean local noon occured at 11:59 am. This calculation can then be compared with a clock set to Greenwich Mean Time to determine our longitude.

2. Project: Curvature of surfaces

Not surprisingly, the shape of the earth affects the results of a geodetic survey of the earth's surface or a portion of it. But it took the genius of the German mathematician, Carl Friedrich Gauss (1777–1855), to imagine the situation in reverse. Assigned in 1818 to head a geodetic survey of the German region of Hannover, Gauss realized that it ought to be possible to use the results of the survey to *determine* the earth's shape. Indeed, one ought to be able to make conclusions about the shape of any surface just by conducting a geodetic survey of it. In this project, you will conduct a miniature "survey" of some selected surfaces and explore how the results compare to those of a similar survey conducted on a flat surface. The results will give us information about our ability or inability to make accurate, true-to-scale flat maps of the different surfaces.

Step 0. Collect at least four objects with interesting surfaces. For instance, a cylindrical food container; a cone for marking a playing field; balls of a couple of different sizes; an inflatable swimming-pool toy in the shape of a ring. You will need to draw on these surfaces, so it is easiest if the surfaces are fairly smooth.

Step 1. On each of the surfaces you have collected, select a base point and then mark off the "circle" of points on the surface which are at a distance of 8 centimeters from the base, where *the distance is measured along the surface.* Measure the perimeter of the "circle" obtained. **Warning: This is not easy!** For instance, you might try using a piece of string that is 8 centimeters long to mark off points on the surface that are 8 centimeters from the base point. Then use a tool such as a "route roller" (available at map or travel shops) to measure the perimeter. Or make a paper disc of radius 8 centimeters and cut some slits in the edge. Lay the disc on the surface so that it conforms to the surface. Some of the slits may overlap when you do this; or the slits may spread apart. Mark the perimeter of the disc as it lies on the surface and measure this perimeter.

Step 2. Repeat Step 1 using the same base point but a distance of 4 centimeters to obtain the "circle". **Record your results.**

Step 3. On a true-to-scale flat map of any of the surfaces you selected, a "circle" of radius $r = 8$ centimeters (measured along the surface) centered at the base point you chose would be shown as a Euclidean circle of radius $r = 8$ centimeters in the plane of the map. The circumference of this circle in the plane of the map would be $2\pi r = 16\pi$ centimeters, or approximately 50.2 centimeters. Similarly, a "circle" of radius $r = 4$ centimeters (measured along the surface) centered at the base point on the surface would be shown as a circle of radius $r = 4$ centimeters in the plane of the map. The circumference of this circle in the plane of the map would be $2\pi r = 8\pi$ centimeters, or approximately 25.1 centimeters.

Roughly speaking, Gauss used the discrepancy between the measured perimeter of a "circle" of radius r on the surface and the quantity $2\pi r$ (which would be the perimeter on a flat surface) as a measure of what we today call the *Gaussian curvature* of the surface at the base point. The Gaussian curvature at a point is *positive* if the measured perimeter is less than $2\pi r$ and *negative* if the measurement is greater than $2\pi r$. A flat surface has zero curvature. Thus, a sphere has positive Gaussian curvature, a saddle shape has negative Gaussian curvature, and a plane has a Gaussian curvature of 0. (The actual amount of Gaussian curvature is obtained by evaluating the expression

$$\left(\frac{3}{\pi}\right)\frac{2\pi r - \text{(measured perimeter)}}{r^3}$$

for small values of r; taking $r = 4$ centimeters should give a better estimate than $r = 8$ centimeters.)

Using this information, compare the perimeters you measured for the radii $r = 8$ centimeters and $r = 4$ centimeters to the quantities 50.2 centimeters and 25.1 centimeters (which would be the perimeters on a flat surface) and decide whether each surface you examined has positive, negative, or zero Gaussian curvature. Make a rough estimate of the actual amount of Gaussian curvature at the base point you selected for each surface.

Step 4. A final question: *Why can we not make a true-to-scale flat map of a surface that has non-zero Gaussian curvature?*

3. Project: World map design

In this set of related laboratory projects, you will construct several maps of the world based on projections discussed in the text. Both hand-drawing and computer techniques will be employed. The design process consists essentially of two parts. First, the map's base grid of parallels and meridians is created. Then drafting techniques are used to transfer continent boundaries onto the map from a reference globe. Cartographers refer to this second step as *compilation*.

3.1. Creating the base grid. You may choose any projection you like, of course, but here we will indicate how to create the base grids for an azimuthal equidistant projeciton, the Mollweide projection, and Mercator's map.

Azimuthal equidistant projection

The base grid for this map can be created completely by hand – about as low-tech as it can get! You will need: (i) a compass (the sort used for drawing circles, not for determining direction); (ii) a ruler; (iii) a protractor; (iv) a drawing pencil (a harder lead is better); (v) a sheet of plain paper. Follow these steps to create the base grid for an azimuthal equidistant projection having a polar perspective and showing one hemisphere. The map will show latitudes in 10 degree increments and longitudes in 20 degree increments.

(a) Select a point in the center of the sheet of paper. This will represent either the north or south pole. With the tip of the compass at this central point, draw nine (9) concentric circles. To produce an azimuthal equidistant projection, **it is essential that you increment the radius of each circle by the same amount to get the radius for the next larger circle.** These circles represent the parallels of latitude. The largest circle represents the equator.

(b) Next use the protractor and a straight edge to mark off radial line seg-
    ments, at increments of 20 degrees, emanating from the central point and
    extending out to the largest of the concentric circles. These lines represent
    the meridians. You now have a complete base grid for your map!

Mollweide projection

To create a computer-generated base grid for the Mollweide projection, you will
need, not surprisingly, a computer with a computer graphics package (e.g., Maple
or Mathematica).

(a) As detailed in the text, the parallel at latitude ϕ, for $\phi \geq 0$, is shown on
    the map as a segment of the horizontal line $y = h = \sqrt{2}\sin(t)$, where t
    satisfies the equation $\pi \sin(\phi) = 2t + \sin(2t)$. This equation does not yield
    a simple expression for t in terms of ϕ. Instead, plug in the selected values
    of ϕ that you wish to depict and solve for the corresponding value of t in
    each case. Then compute the value of h. Table 1 in Chapter 8 shows the
    results for certain values of ϕ.

(b) The meridians at longitude θ and $-\theta$, when taken together, are depicted
    on the map as the ellipse with equation $\pi^2 x^2 + 4\theta^2 y^2 = 8\theta^2$. Again, select
    the values of θ that you wish to depict and plot the appropriate ellipses.

(c) A detail that must be attended to in order to produce a nice looking
    base grid is the determination of the correct lengths for the horizontal
    segments you found in part (a). This amounts to computing the points
    of intersection of the horizontal line $y = h$ and the outer boundary of the
    map, the ellipse $x^2 + 4y^2 = 8$. This occurs when $x = \pm 2\sqrt{2 - h^2}$, so that
    is the correct interval over which to plot the parallel.

(d) Once you have successfully plotted all of these elements in a single picture,
    print out a full-page copy of the base grid. You may wish to orient the
    paper in your printer to "landscape" mode as the grid for this projection
    is wider than it is tall.

One example of Maple code that can be used to generate this grid is given in
Appendix C.

Mercator's map

To create a computer-generated base grid for Mercator's map, you will need, of
course, a computer with a computer graphics package (e.g., Maple or Mathematica).

As described in the text, this grid consists of vertical lines of the form $x = \theta$
and horizontal line segments of length 2π of the form $y = \pm \ln|\sec(\phi) + \tan(\phi)|$,
where θ denotes the longtitude and ϕ is the latitude. Plug in the values for ϕ that
you want to depict in your grid to compute the corresponding y values. If you
don't have a computer, you could actually compute these y values on a calculator,
then use a ruler to draw horizontal lines at approximately the correct heights. Note
that it is not possible to plot the poles on the base grid, so cut the meridians off at
75 or 80 degrees north and south. Once you have successfully plotted all of these
elements in a single picture, **print out a full-page copy** of the base grid.

3.2. Map compilation. The following steps explain how to create a finished map from the base grid. You will need (i) a sheet of drafting paper; (ii) drafting tape; (iii) a drawing pencil (a harder lead is better); (iv) a straight edge; (v) a globe for reference.

(1) **Prepare the drafting paper.** Properly tape your base grid (prepared in the previous step) to a table or other flat surface suitable for drawing. (In order to reduce the tape's stickiness a bit, it helps to stick the tape to your clothes before using it on paper.) Use your drawing pencil and a straight edge to draw a pair of crossed lines (called a *registration mark*) in three of the four corners of the base grid sheet. Place the sheet of drafting paper over the base grid and tape it down. Then, using straight edge and drafting pencil, trace the registration marks onto your drafting paper. These registration marks will allow you to properly align all of your work for this lab. Write your name, in pencil, in the lower right-hand corner of your drafting paper.

(2) **Some choices to be made.**

    (a) For the azimuthal equidistant projection, decide which hemisphere (northern or southern) you will map. The pole for the hemisphere you choose will be the center of your map. Once you have chosen the hemisphere to be mapped, you must select a meridian (one of the lines radiating from the center on your base grid) to serve as the prime meridian for your map. You may choose any one of these radiating lines to label as the prime meridian. It does not matter which. It will probably be helpful at this point to *lightly* label *in pencil* the correct longitude for each meridian. Make sure that you get the east/west orientation correct! Believe it or not, reversing east and west is one of the most common mistakes made when carrying out this project.

    (b) For the Mollweide projection or Mercator's map, there is no particular reason why the north pole must be at the top, so you may choose to have the south pole at the top of your map. Second, the prime meridian need not be the central meridian on the map, so you must select which meridian you wish to have at the center. For instance, a map centered on the meridian 60°W would have South America at the center. It may be helpful at this point to *lightly* label *in pencil* the correct longitude for each meridian.

(3) **The Like Squares Method.** The "like squares method" is a low-tech method for sketching continent boundaries on a map. It is based on transferring information from a globe to a map one "square" at a time. Here is how it works.

    For each "square" on the map's base grid, bounded by two lines of latitude and two lines of longitude, locate the same "square" on the globe, bounded by the *same lines* of latitude and longitude. Then, as accurately as you can, copy the land boundaries you find in that "square" of the globe onto the corresponding "square" on your map. You should work on a square-by-square basis, taking care to work with regions that are bounded by the same lines of latitude and longitude on your reference globe and on your base grid. Using different regions on the globe and on the base grid is one of the main mistakes that can occur in compiling the map. (For instance, if the globe shows meridians every 10 degrees and your base grid only shows meridians every 20 degrees, then one "square" on the base grid will correspond to two smaller squares on the globe.)

(4) **Add continent boundaries.** Make sure that you have determined the appropriate value (how many degrees north or south, east or west) for each line of latitude and longitude on the base grid. Check once again that you have the correct east/west orientation. Then use the like squares method just described to compile a map of the appropriate land boundaries. **Do not trace the base grid itself onto the drafting paper.** At this point, the base grid sheet is serving only to guide your work in compiling land boundaries.

(5) **Finishing touches.**

  (a) Carefully trace onto the drafting paper *only* those portions of the base grid that lie within water areas. This will result in the land areas standing out visually against the background of the oceans. Land will represent "figure" (foreground) and the oceans will represent "ground" (background) in your illustration.

  (b) Give your completed map a creative title and include this centered below your map.

Be patient. Be careful. This takes time.

Draw lightly. You may need to erase and correct your line work.

This is not necessarily an easy task.

If you have problems, or are unsure, seek help.

Enjoy!

4. Project: Navigation

In this project, you will use a gnomonic projection and a Mercator projection together to solve a navigation problem.

You will need: (i) a straight edge; (ii) a protractor; (iii) a drawing pencil; (iv) a globe or atlas; (v) base grids for two maps – one for a gnomonic projection, the other for a Mercator projection – each covering the region over which the navigating is to take place.

(1) On the gnomonic grid, mark two locations, A and B, that lie in the common region covered by the grids. (You may choose any two places of interest to you as long as they are both contained within the common region of both grids.) Connect these locations with a straight line segment. On this line segment, mark at least three (3) additional points that lie between A and B.

(2) Next, on the Mercator grid, mark the locations of A, B, and the intermediate points you chose. Connect these points "dot-to-dot" using straight line segments.

(3) Use a protractor to determine the compass bearing (measured as an angle relative to the direction north) of each of the line segments you just drew.

(4) Describe in words, using correct grammar and complete sentences, what each of the two paths you have drawn represents and discuss why an airline pilot might be interested in the map work you have just completed.

Portraits of the earth: How the maps in this book were produced

1. The Maptools package

In 2001, in order to facilitate the process of world map generation, Vincent Costanzo, then a student at Villanova University, developed a set of commands and procedures within the framework of the computer algebra system Maple that he named the "Maptools" package. Available to the public via the Maplesoft applications center site on the World Wide Web, the Maptools package allows the user to enter either two- or three-dimensional Cartesian coordinates for a map projection. The package then incorporates a database of coastal points that Costanzo adapted from the United States Geological Survey to produce a portrait of both the base grid for the map and the continental coastlines. Various options in the package allow the user to customize the portrait, including restricting the ranges of longitude and latitude, for example.

To use the Maptools package, one must first download the library of procedures and the database of coastal points from the Maplesoft applications center. These must then be loaded into the Maple worksheet itself to create a map.

For example, to create the stereographic projection of the southern hemisphere shown in Figure 1 of Chapter 7, the first step is to determine the Cartesian coordinates for the projection. Since $x = r(\phi)\cos(\theta)$ and $y = r(\phi)\sin(\theta)$, the projection is determined by the mapping

$$(\theta, \phi) \rightarrow \left(\frac{2\cos(\phi)\cos(\theta)}{1 - \sin(\phi)}, \frac{2\cos(\phi)\sin(\theta)}{1 - \sin(\phi)} \right).$$

The longitude and latitude restrictions for the southern hemisphere are $-\pi \leq \theta \leq \pi$ and $-\pi/2 \leq \phi \leq 0$. The resulting Maple worksheet is as follows, where v and u have been used in place of θ and ϕ, respectively. In the worksheet, "c:/mplfiles/vincemaps" is the path on my local machine to the downloaded Maptools package. Your path name will likely be different. The name "stereocoords" identifies the formula for the projection while the command "mapplot" applies the formula to produce the map itself.

```
>  with(plots):  libname:="c:/mplfiles/vincemaps",libname;
with(maptools):  load("c:/mplfiles/vincemaps/complete.m");
>  stereocoords:=(v,u)->(evalf(2*cos(u)*cos(v)/(1-sin(u))),
evalf(2*cos(u )*sin(v)/(1-sin(u))));
>  mapplot(stereocoords,lonlat=[-Pi..Pi,-Pi/2..0]);
```

1.1. Variations. Some map projections are not defined by simple formulas for the map's coordinates. With the Mollweide projection, for instance, the point on the globe with longitude θ and latitude ϕ is mapped to a point whose y-coordinate is equal to $h(\phi) = \sqrt{2}\sin(t)$ where t, in turn, satisfies the condition $\pi\sin(\phi) = 2t + \sin(2t)$ as in equation (23). The x-coordinate of the image point is then given by $x = (2\theta/\pi)\sqrt{2 - h(\phi)^2}$ in accordance with equation (30). Thus, for each coastal point we wish to plot on our map, we must numerically solve for $h(\phi)$ and then use this value to compute the x-coordinate. The Maptools package does not accept this as a simple formula. We can, however, extract some of the essence of the Maptools package and work with the list of coastal points one point at a time. In this way we can generate a list of image points to be mapped. Maple's "pointplot" command can then be used to generate a map of the coastal points.

It is fairly straightforward to generate the base grid of the Mollweide projection as this consists of certain horizontal line segments and ellipses for a relatively small number of selected values of ϕ and θ. Maple's "display" command then enables us to portray both the base grid and the image of the coastal points in the same picture. The Maple worksheet that was used to generate the Mollweide projection in Figure 6 of Chapter 8 is given here. Again, v and u have been used in place of θ and ϕ. The name "imagecoast" is for the list of images of the coastal points, while "coastmap" is the plot of this list. The meridians, in twenty degree increments, and the parallels, in ten degree increments, are represented by "merplot" and "latplot", respectively. The prime meridian, which is portrayed not as an ellipse but as a vertical line segment, is constructed separately as "primemeridian".

```
>  with(plots):   libname:="c:/mplfiles/vincemaps",libname;
with(maptools):   load("c:/mplfiles/vincemaps/complete.m");

>  t1:=(v,u)->fsolve(Pi*sin(u)=2*t+sin(2*t),t=-Pi/2..Pi/2);

>  h1:=(v,u)->evalf(sqrt(2)*sin(t1(v,u)));

>  xmapcoord:=(v,u)->evalf(2*(v/Pi)*sqrt(2-h1(v,u)^2));

>  ymapcoord:=(v,u)->h1(v,u);

>  imagecoast:=[seq([xmapcoord(coastal_points[ii][1],
coastal_points[ii][2]),ymapcoord(coastal_points[ii][1],
coastal_points[ii][2])],ii=1..nops(coastal_points))]:

>  coastmap:=pointplot(imagecoast,scaling=constrained,axes=none,
symbol=point,color=black):

>  primemeridian:=implicitplot(x=0,x=-2*sqrt(2)..2*sqrt(2),
y=-sqrt(2)..sqrt(2),axes=none,scaling=constrained,color=grey):

>  merplot:=seq(implicitplot(81*x^2/(8*n^2)+y^2/2=1,
x=-2*sqrt(2)..2*sqrt (2),y=-sqrt(2)..sqrt(2),axes=none,
scaling=constrained,color=grey),n=1..9):

>  latplot:=seq(plot({h1(0,n*Pi/18),-h1(0,n*Pi/18)},
x=-2*sqrt(2-h1(0,n*Pi/18)^2)..2*sqrt(2-h1(0,n*Pi/18)^2),
axes=none,scaling=constrained,color=grey),n=0..8):

>  display({primemeridian,merplot,latplot,coastmap});
```

The globular projection of al-Biruni, shown in Figure 5 of Chapter 7, was created in a similar manner.

2. Multi-component displays

Several pictures in the book, such as Figure 1 in Chapter 4, Figure 2 in Chapter 9, and Figure 2 in Chapter 10, consist of a globe or map together with other elements. Each component of the figure is created separately and Maple's "display" command is then used to display them all in the same window.

For instance, Figure 1 in Chapter 4 shows part of the globe together with the great circle route from London to Beijing. The globe was created with Maptools using the "mapplot3d" command and the standard conversion of longitude and latitude into three-dimensional Cartesian coordinates. Thus,

```
> globecoords:=(v,u)->(evalf(cos(v)*cos(u)),
evalf(sin(v)*cos(u)),evalf( sin(u)));
> p1:=mapplot3d(globecoords):
```

The plot of the globe has been named 'p1'. The Maptools package must be loaded first in order for the command "mapplot3d" to make sense.

The great circle arc that connects London to Beijing is a curve in space and, as such, can be parametrized. (You may have worked out how to do this in exercise 6 of Chapter 4.) We need to express each city as a vector on a globe of radius 1 and compute the angle, call it $\theta 1$, between these vectors as described in equation (5). Then apply Maple's "spacecurve" command to the parametrized arc to get the picture, called 'LotoBe' here. Here is one way to do this.

```
> u1:=convert(51.5*degrees, radians); v1:=0;
> u2:=convert(40*degrees, radians);
v2:=convert(116.35*degrees, radians);
> theta1:=arccos(cos(u1)*cos(u2)*cos(v1-v2)+sin(u1)*sin(u2));
> Lo:=vector([cos(v1)*cos(u1),sin (v1)*cos(u1),sin(u1)]);
Be:=vector([cos(v2)*cos(u2),sin(v2)*cos(u2),sin(u2)]);
> LotoBe:=spacecurve([(cos(t)-sin(t)*cot(theta1))*Lo[1]+
(sin(t)/sin (theta1))*Be[1],(cos(t)-sin(t)*cot(theta1))*Lo[2]+
(sin(t)/sin(theta1)) *Be[2],(cos(t)-sin(t)*cot(theta1))*Lo[3]+
(sin(t)/sin(theta1))*Be[3]],
t=0..theta1,color=black,thickness=2):
```

Finally, the command

```
> display({p1,LotoBe},orientation=[50,60]);
```

puts the globe and the great circle arc together in a display.

The other figures are left as exercises. Challenge yourself!

Bibliography

[1] Berggren, J. Lennart, and Alexander Jones. *Ptolemy's Geography*. Princeton: Princeton University Press. 2000.

[2] Bosowski, Elaine F., and Timothy G. Feeman. *The use of scale factors in map analysis: an elementary approach*, Cartographica, Volume 34, No.4, 1997, pp. 35–44.

[3] Bugayevskiy, Lev M., and John P. Snyder. *Map Projections: A Reference Manual*. London: Taylor and Francis. 1995.

[4] Bühler, W. K. *Gauss: A Biographical Study*. Berlin, Heidelberg, New York: Springer-verlag. 1981.

[5] Campbell, John. *Map Use and Analysis*, 2nd edition. Dubuque: Wm. C. Brown. 1993.

[6] Cotter, Charles H. *The Astronomical and Mathematical Foundations of Geography*. New York: American Elsevier. 1966.

[7] Coulson, Michael R. C. *In praise of dot maps*, International Yearbook of Cartography, volume 30 (1990), pp. 61–91.

[8] Deetz, Charles H., and Oscar S. Adams. *Elements of Map Projection*. New York: Greenwood Press. 1969.

[9] Dent, Borden D. *Cartography*, 4th edition. Dubuque: Wm. C. Brown. 1996.

[10] Espenshade, Edward B., Jr., et al., ed. *Goode's World Atlas*, 19th edition. Rand McNally. 1995.

[11] Feeman, Timothy G. *Equal area world maps: A case study*, SIAM Review, Volume 42, No.1, 2000, pp. 109–114.

[12] Feeman, Timothy G. *Conformality, the exponential function, and world map projections*, College Math. Journal, Volume 32, No. 5, 2001, pp. 334–342.

[13] Harley, J. B. *The New Nature of Maps*, Baltimore: The Johns Hopkins University Press. 2001.

[14] Kaiser, Ward L. *A New View of the World*, New York: Friendship Press. 1993.

[15] Keller, C. Peter. Towards an Introductory Cartographic Curriculum for the 21st Century. *Cartographica* Volume 33, Number 3, 1996, pp. 45–53.

[16] Krantz, Steven G. *Conformal mappings*, American Scientist, 87 (1999), 436–445.

[17] MacEachren, Alan M. *How Maps Work: Representation, Visualization, and Design*. New York: Guilford Press. 1995.

[18] McCleary, John. *Geometry From a Differentiable Viewpoint*. Cambridge: Cambridge University Press. 1994.

[19] Monmonier, Mark. *Drawing the Line: Tales of Maps and Cartocontroversy*. New York: Henry Holt. 1995.

[20] Monmonier, Mark. *How To Lie With Maps*. Chicago: University of Chicago Press. 1991.

[21] Osserman, Robert. *Poetry of the Universe*. New York: Anchor Books. 1995.

[22] Pearson, Frederick, II. *Map Projections: Theory and Applications*. Boca Raton: CRC Press. 1990.

[23] Robinson, Arthur H., et al. *Elements of Cartography*. New York: John Wiley. 1995.

[24] Snyder, John P. *Flattening the Earth: Two Thousand Years of Map Projections*. Chicago: University of Chicago Press. 1993.

[25] Sobel, Dava. *Longitude*. New York: Walker and Co. 1995.

[26] Struik, Dirk J. *A Concise History of Mathematics*, 4th revised edition. New York: Dover. 1987.

[27] Toomer, G. J. *Ptolemy's Almagest*. Princeton: Princeton University Press. 1998.

[28] Wilford, John Noble. *The Mapmakers*. New York: Alfred A. Knopf. 2000.

[29] Wood, Denis. *The Power of Maps*. New York: Guilford Press. 1992.

Index

Titles in This Series